MEMBRANE TECHNOLOGIES FOR WATER TREATMENT: REMOVAL OF TOXIC TRACE ELEMENTS WITH EMPHASIS ON ARSENIC, FLUORIDE AND URANIUM

Sustainable Water Developments
Resources, Management, Treatment, Efficiency and Reuse

Series Editor

Jochen Bundschuh
University of Southern Queensland (USQ), Toowoomba, Australia
Royal Institute of Technology (KTH), Stockholm, Sweden

ISSN: 2373-7506

Volume 1

Membrane Technologies for Water Treatment: Removal of Toxic Trace Elements with Emphasis on Arsenic, Fluoride and Uranium

Editors

Alberto Figoli

Institute of Membrane Technology, ITM-CNR, Rende (CS), Italy

Jan Hoinkis

Karlsruhe University of Applied Sciences, Institute of Applied Research, Karlsruhe, Germany

Jochen Bundschuh

Deputy Vice Chancellor's Office (Research and Innovation) & Faculty of Health, Engineering and Sciences, Toowoomba, Queensland, Australia & Royal Institute of Technology (KTH), Stockholm, Sweden

CRC Press is an imprint of the
Taylor & Francis Group, an **informa** business
A BALKEMA BOOK

Cover photo

The upper photo shows one of the largest reverse osmosis plants in South America, which treats arsenic-laden water coming from the boreholes in Virrey Del Pino in La Matanza, Buenos Aires, Argentina by means of the reverse osmosis system. It produces 47,040 m^3 of drinking water per day for 400,000 inhabitants.

The other photo shows the tap of an autonomous desalination plant Mörk Desalin™ for drinking water production in developing countries which has been put on the market by the German company Mörk Water Solutions, Leonberg.

CRC Press
Taylor & Francis Group
6000 Broken Sound Parkway NW, Suite 300
Boca Raton, FL 33487-2742
First issued in paperback 2021

© 2016 by Taylor & Francis Group, LLC
CRC Press is an imprint of Taylor & Francis Group, an Informa business
Typeset by MPS Limited, Chennai, India

No claim to original U.S. Government works

ISBN-13: 978-1-138-02720-6 (hbk)
ISBN-13: 978-1-138-61193-1 (pbk)

This book contains information obtained from authentic and highly regarded sources. Reasonable efforts have been made to publish reliable data and information, but the author and publisher cannot assume responsibility for the validity of all materials or the consequences of their use. The authors and publishers have attempted to trace the copyright holders of all material reproduced in this publication and apologize to copyright holders if permission to publish in this form has not been obtained. If any copyright material has not been acknowledged please write and let us know so we may rectify in any future reprint.

Except as permitted under U.S. Copyright Law, no part of this book may be reprinted, reproduced, transmitted, or utilized in any form by any electronic, mechanical, or other means, now known or hereafter invented, including photocopying, microfilming, and recording, or in any information storage or retrieval system, without written permission from the publishers.

For permission to photocopy or use material electronically from this work, please access www.copyright.com (http://www.copyright.com/) or contact the Copyright Clearance Center, Inc. (CCC), 222 Rosewood Drive, Danvers, MA 01923, 978-750-8400. CCC is a not-for-profit organization that provides licenses and registration for a variety of users. For organizations that have been granted a photocopy license by the CCC, a separate system of payment has been arranged.

Trademark Notice: Product or corporate names may be trademarks or registered trade- marks, and are used only for identification and explanation without intent to infringe.

Publisher's Note

The publisher has gone to great lengths to ensure the quality of this reprint but points out that some imperfections in the original copies may be apparent.

Library of Congress Cataloging-in-Publication Data

Names: Figoli, Alberto, editor. | Hoinkis, Jan, editor. | Bundschuh, Jochen, editor.

Title: Membrane technologies for water treatment : removal of toxic trace elements with emphasis on arsenic, fluoride and uranium / editors, Alberto Figoli, Institute of Membrane Technology, ITM-CNR, Rende (CS), Italy, Jan Hoinkis, Karlsruhe University of Applied Sciences, Institute of Applied Research, Karlsruhe, Germany, Jochen Bundschuh, Deputy Vice Chancellor's Office (Research and Innovation) & Faculty of Health, Engineering and Sciences, Toowoomba, Queensland, Australia & Royal Institute of Technology (KTH), Stockholm, Sweden.

Description: Leiden, The Netherlands; Boca Raton: CRC Press/Balkema, [2016] | Series: Sustainable water developments; volume 1 | Includes bibliographical references and index.

Identifiers: LCCN 2015050193 (print) | LCCN 2016004943 (ebook) | ISBN 9781138027206 (hardcover : alk. paper) | ISBN 9781315735238 (eBook PDF) | ISBN 9781315735238 (ebook)

Subjects: LCSH: Water—Purification—Membrane filtration. | Water—Purification—Arsenic removal. | Arsenic—Environmental aspects. | Uranium—Environmental aspects. | Fluoride—Environmental aspects.

Classification: LCC TD442.5 .M4586 2016 (print) | LCC TD442.5 (ebook) | DDC 628.1/64—dc23

LC record available at http://lccn.loc.gov/2015050193

Visit the Taylor & Francis Web site at
http://www.taylorandfrancis.com

and the CRC Press Web site at
http://www.crcpress.com

About the book series

Augmentation of freshwater supply and better sanitation are two of the world's most pressing challenges. However, such improvements must be done economically in an environmentally and societally sustainable way.

Increasingly, groundwater – the source that is much larger than surface water and which provides a stable supply through all the seasons – is used for freshwater supply, which is exploited from ever-deeper groundwater resources. However, the availability of groundwater in sufficient quantity and good quality is severely impacted by the increased water demand for industrial production, cooling in energy production, public water supply and in particular agricultural use, which at present consumes on a global scale about 70% of the exploited freshwater resources. In addition, climate change may have a positive or negative impact on freshwater availability, but which one is presently unknown. These developments result in a continuously increasing water stress, as has already been observed in several world regions and which has adverse implications for the security of food, water and energy supplies, the severity of which will further increase in future. This demands case-specific mitigation and adaptation pathways, which require a better assessment and understanding of surface water and groundwater systems and how they interact with a view to improve their protection and their effective and sustainable management.

With the current and anticipated increased future freshwater demand, it is increasingly difficult to sustain freshwater supply security without producing freshwater from contaminated, brackish or saline water and reusing agricultural, industrial, and municipal wastewater after adequate treatment, which extends the life cycle of water and is beneficial not only to the environment but also leads to cost reduction. Water treatment, particularly desalination, requires large amounts of energy, making energy-efficient options and use of renewable energies important. The technologies, which can either be sophisticated or simple, use physical, chemical and biological processes for water and wastewater treatment, to produce freshwater of a desired quality. Both industrial-scale approaches and smaller-scale applications are important but need a different technological approach. In particular, low-tech, cost-effective, but at the same time sustainable water and wastewater treatment systems, such as artificial wetlands or wastewater gardens, are options suitable for many small-scale applications. Technological improvements and finding new approaches to conventional technologies (e.g. those of seawater desalination), and development of innovative processes, approaches, and methods to improve water and wastewater treatment and sanitation are needed. Improving economic, environmental and societal sustainability needs research and development to improve process design, operation, performance, automation and management of water and wastewater systems considering aims, and local conditions.

In all freshwater consuming sectors, the increasing water scarcity and correspondingly increasing costs of freshwater, calls for a shift towards more water efficiency and water savings. In the industrial and agricultural sector, it also includes the development of technologies that reduce contamination of freshwater resources, e.g. through development of a chemical-free agriculture. In the domestic sector, there are plenty of options for freshwater saving and improving efficiency such as water-efficient toilets, water-free toilets, or on-site recycling for uses such as toilet flushing, which alone could provide an estimated 30% reduction in water use for the average household. As already mentioned, in all water-consuming sectors, the recycling and reuse of the respective wastewater can provide an important freshwater source. However, the rate at which these water efficient technologies and water-saving applications are developed and adopted depends on the behavior of individual consumers and requires favorable political, policy and financial conditions.

Due to the interdependency of water and energy (water-energy nexus); i.e. water production needs energy (e.g. for groundwater pumping) and energy generation needs water (e.g. for cooling), the management of both commodities should be more coordinated. This requires integrated energy and water planning, i.e. management of both commodities in a well-coordinated form rather than managing water and energy separately as is routine at present. Only such integrated management allows reducing trade-offs between water and energy use.

However, water is not just linked to energy, but must be considered within the whole of the water-energy-food-ecosystem-climate nexus. This requires consideration of what a planned water development requires from the other sectors or how it affects – positively or negatively – the other sectors. Such integrated management of water and the other interlinked resources can implement synergies, reduce trade-offs, optimize resources use and management efficiency, all in all improving security of water, energy, and food security and contributing to protection of ecosystems and climate. Corresponding actions, policies and regulations that support such integral approaches, as well as corresponding research, training and teaching are necessary for their implementation.

The fact that in many developing and transition countries women are disproportionately disadvantaged by water and sanitation limitation requires special attention to this aspect in these countries. Women (including schoolgirls) often spend several hours a day fetching water. This time could be much better used for attending school or working to improve knowledge and skills as well as to generate income and so to reduce gender inequality and poverty. Absence of in-door sanitary facilities exposes women to potential harassment. Moreover, missing single-sex sanitation facilities in schools and absence of clean water contributes to diseases. This is why women and girls are a critical factor in solving water and sanitation problems in these countries and necessitates that men and women work side by side to address the water and wastewater related operations for improvement of economic, social and sustainable freshwater provision and sanitation.

Individual volumes published in the series span the wide spectrum between research, development and practice in the topic of freshwater and related areas such as gender and social aspects as well as policy, regulatory, legal and economic aspects of water. The series covers all fields and facets in optimal approaches to the:

- Assessment, protection, development and sustainable management of groundwater and surface water resources thereby optimizing their use.
- Improvement of human access to water resources in adequate quantity and good quality.
- Meeting of the increasing demand for drinking water, and irrigation water needed for food and energy security, protection of ecosystems and climate and contribution to a socially and economically sound human development.
- Treatment of water and wastewater also including its reuse.
- Implementation of water efficient technologies and water saving measures.

A key goal of the series is to include all countries of the globe in jointly addressing the challenges of water security and sanitation. Therefore, we aim for a balanced selection of authors and editors originating from developing and developed countries as well as for gender equality. This will help society to provide access to freshwater resources in adequate quantity and of good quality, meeting the increasing demand for drinking water, domestic water and irrigation water needed for food security while contributing to socially and economically sound development.

This book series aims to become a state-of-the-art resource for a broad group of readers including professionals, academics and students dealing with ground and surface water resources, their assessment, exploitation and management as well as the water and wastewater industry. This comprises especially hydrogeologists, hydrologists, water resources engineers, wastewater engineers, chemical engineers and environmental engineers and scientists.

The book series provides a source of valuable information on surface water but especially on aquifers and groundwater resources in all their facets. As such, it covers not only the scientific and technical aspects but also environmental, legal, policy, economic, social, and gender aspects of groundwater resources management. Without departing from the larger framework of integrated groundwater resources management, the topics are centered on water, solute and heat transport in aquifers, hydrogeochemical processes in aquifers, contamination, protection, resources assessment and use.

The book series constitutes an information source and facilitator for the transfer of knowledge, both for small communities with decentralized water supply and sanitation as well as large industries that employ hundreds or thousands of professionals in countries worldwide, working in the different fields of freshwater production, wastewater treatment and water reuse as well as those concerned with water efficient technologies and water saving measures. In contrast to many other industries, suffering from the global economic downturn, water and wastewater industries are rapidly growing sectors providing significant opportunities for investments. This applies especially to those using sustainable water and wastewater technologies, which are increasingly favored. The series is also aimed at communities, manufacturers and consultants as well as a range of stakeholders and professionals from governmental and non-governmental organizations, international funding agencies, public health, policy, regulating and other relevant institutions, and the broader public. It is designed to increase awareness of water resources protection and understanding of sustainable water and wastewater solutions including the promotion of water and wastewater reuse and water savings.

By consolidating international research and technical results, the objective of this book series is to focus on practical solutions in better understanding groundwater and surface water systems, the implementation of sustainable water and wastewater treatment and water reuse and the implementation of water efficient technologies and water saving measures. Failing to improve and move forward would have serious social, environmental and economic impacts on a global scale.

The book series includes books authored and edited by world-renowned scientists and engineers and by leading authorities in economics and politics. Women are particularly encouraged to contribute, either as author or editor.

Jochen Bundschuh
(Series Editor)

Editorial board

Blanca Jiménez Cisneros
Director of the Division of Water Sciences, Secretary of the International Hydrological Programme (IHP), UNESCO, Paris, France

Glen T. Daigger
Immediate Past President, International Water Association (IWA), IWA Board Member, Senior Vice President and Chief Technology Officer CH2M HIL, Englewood, CO, USA

Anthony Fane
Director, Singapore Membrane Technology Centre (SMTC), Nanyang Technological University (NTU), Singapore

Carlos Fernandez-Jauregui
Director, International Water Chair-EUPLA, Director, Water Assessment & Advisory Global Network (WASA-GN), Madrid, Spain

ADVISORY EDITORIAL BOARD

AFGHANISTAN

Naim Eqrar (water resources; water quality; hydrology; hydrogeology; environmental awareness), Full Member of the Secretariat of Supreme Council of Water and Land of Afghanistan (SCWLA); Full Member of the National Hydrological Committee of Afghanistan (NHCA) & Geoscience Faculty, Kabul University, Kabul

ALGERIA

Hacene Mahmoudi (renewable energy for desalination and water treatment; membrane technologies), Faculty of Sciences, Hassiba Ben Bouali University of Chlef (UHBC), Chlef

ANGOLA

Helder de Sousa Andrade (geomorphology; impact of land use on water resources (water courses, rivers and lakes); climate variability impacts on agriculture), Department of Geology, Faculty of Sciences, Agostinho Neto University (UAN), Luanda

ARGENTINA

Alicia Fernández Cirelli (water quality; aquatic ecosystems; aquatic humic substances; heavy metal pollution; use of macrophytes as biosorbents for simultaneous removal of heavy metals; biotransfer of arsenic from water to the food chain), Director, Center for Transdisciplinary Studies on Water Resources, Faculty of Veterinary Sciences, University of Buenos Aires (UBA) & National Scientific and Technical Research Council (CONICET), Buenos Aires

Carlos Schulz (hydrogeology; groundwater management planning; water chemistry; groundwater-surface water interactions; trace element contaminants), President IAH Chapter Argentina; Faculty of Exact and Natural Sciences, National University of La Pampa (UNLPam), Santa Rosa, La Pampa

ARMENIA

Vilik Sargsyan (hydrology; water resources; assessment of environmental flows; transboundary rivers), Yerevan National University of Architecture and Construction of Armenia, Yerevan

ARUBA

Filomeno A. Marchena (seawater desalination for the production of drinking and industrial water: improving excellence performance and efficiency, improvement through theoretical and practical research and knowledge transfer, sustainable green desalination and energy technologies for Small Islands Development States (SIDS); drinking water conditioning; corrosion inhibition), Chair "Sustainable Water Technology and Management", Faculty of Technical Sciences, University of Curacao Dr. Moises Da Costa Gomez; President and Founder of the Aruba Sustainable Water Education Foundation FESTAS; National Focal Point for the International Hydrological Programme (IHP) of UNESCO; Advisor Sustainable Desalination Technology and Innovation at WEB Aruba NV, Oranjestad

AUSTRALIA

Habib Alehossein (geothermal aquifers; geomechanics; modeling), School of Civil Engineering and Surveying, Faculty of Health, Engineering and Science, University of Southern Queensland (USQ), Toowoomba, QLD

Vasanthadevi Aravinthan (environmental water engineering), School of Civil Engineering and Surveying, Faculty of Health, Sciences and Engineering, University of Southern Queensland (USQ), Toowoomba, QLD

Hui Tong Chua (thermal desalination; low grade heat driven multi-effect distillation), School of Mechanical and Chemical Engineering, University of Western Australia (UWA), Perth, WA

Laszlo Erdei (wastewater treatment; membrane technologies), School of Civil Engineering and Surveying, Faculty of Health, Engineering and Science, University of Southern Queensland (USQ), Toowoomba, QLD

Howard Fallowfield (aquatic microbial ecology; health aspects of water quality (drinking water, wastewater and recreational waters); design, construction, operation and evaluation of the performance of pilot plants for both drinking water and wastewater treatment; application of HRAPs for the treatment of wastewaters and on-site biomass energy production; removal of pharmaceuticals and personal care products from water using biological filters), Health & Environment Group, School of the Environment, Flinders University, Bedford Park, SA

Ashantha Goonetilleke (integrated water resources management; climate change adaptation; stormwater quality; pollutant processes and treatment; water sensitive urban design; water pollution, recycling, security and infrastructure resilience), Science and Engineering Faculty, Queensland University of Technology (QUT), Brisbane, QLD

Stephen Gray (chemical coagulation; membrane distillation; membrane fouling; brine management; solid/liquid separation processes; sustainable water systems; water treatment; silica removal), Director, Institute for Sustainability and Innovation (ISI), Victoria University (VU), Melbourne, VIC

Goen Ho (small and decentralized water supply and wastewater systems), School of Engineering and Information Technology, Murdoch University, Perth, WA

Sandra Kentish (membrane technology), Department of Chemical and Biomolecular Engineering, University of Melbourne, Melbourne, VIC

Steven Kenway (urban metabolism and the water-energy nexus; quantifying water-energy linkages (cause and effect relationships) in households, industry, water utilities and cities),

School of Chemical Engineering, Faculty of Engineering, Architecture and Information Technologies, The University of Queensland (UQ), Brisbane, QLD

Nasreen Islam Khan (water quality assessment; modeling and mapping geogenic contaminants; arsenic in food chain and mitigation; human health risk assessment and mapping; willingness to pay; climate change impact on land use; GIS and remote sensing), Fenner School of Environment and Society, Australian National University (ANU), Canberra, ACT

Ross Kleinschmidt (natural and anthropogenic radioactive matter in groundwater, surface water and soil), Radiation & Nuclear Sciences, Forensic & Scientific Services, Health Support Queensland Department of Health, Queensland Government, Coopers Plains, QLD

Mitchell Laginestra (water and wastewater treatment; odor control; winery wastewater management), GHD Australia, Adelaide, SA

Shahbaz Mushtaq (water economics; water-energy nexus; integrated water resources management, regional economic modeling, managing climate variability and climate change, risk management and adaptation, water and agricultural policies development), Acting Director, International Centre for Applied Climate Science, University of Southern Queensland (USQ), Toowoomba, QLD

Long D. Nghiem (removal of trace contaminants by NF/RO membranes; membrane fouling and autopsy studies; non-potable and indirect potable water reuse; membrane bioreactors; membrane extraction; membrane electrolysis), School of Civil Mining and Environmental Engineering, University of Wollongong (UOW), NSW

Gary Owens (environmental contaminants; environmental risk assessment and remediation; engineered nanoparticles remediation; environmental chemistry and toxicology; metal bioavailability; cost effective in-situ and ex-situ remediation techniques), Division of Information Technology, Engineering and the Environment, Future Industries Institute, University of South Australia (UniSA), Adelaide, SA

Neil Palmer (desalination), CEO, National Centre of Excellence in Desalination Australia, Murdoch University, Perth, WA

Ric Pashley (household wastewater treatment and recycling; thermal and membrane desalination; arsenic removal and water sterilization), School of Physical, Environmental & Mathematical Sciences, University of New South Wales (UNSW), Canberra, ACT

Jehangir F. Punthakey (groundwater flow and transport modeling; seawater intrusion; climate change impacts on water resources; indicators for managing stressed aquifers; groundwater resource management and protection; agricultural water management; groundwater policies), Director, Ecoseal Pty Ltd, Roseville, NSW & Adjunct at the Institute for Land, Water and Society, Wagga Wagga Campus, Charles Sturt University, North Wagga Wagga, NSW

Robert J. Seviour (activated sludge microbiology), Department of Microbiology, School of Life Sciences, Faculty of Science, Technology and Engineering, Latrobe University, Bendigo, VIC

Rodney A. Stewart (water resource management (water end-use analysis, recycled water strategies, least cost demand management solutions); environmental management (water and energy conservation); renewable energy systems (PV systems, battery storage, smart grid); smart asset management; water-energy nexus), Director, Centre for Infrastructure Engineering & Management, Griffith School of Engineering, Griffith University, Gold Coast, QLD

Guangzhi Sun (wastewater treatment; tidal-flow artificial wetlands), Chemical Engineering, School of Engineering, Edith Cowan University (ECU), Perth, WA

Simon Toze (reuse of water in urban environment), CSIRO EcoSciences Precinct – Dutton Park, Dutton Park, QLD

Hao Wang (water and wastewater treatment using activated carbon and graphene-like carbon from biomass; hazardous and nuclear waste stabilization using mineral and alkali activated cement), Leader, Biomass Composites and Green Concrete Programs, Centre of Excellence in Engineered Fibre Composites (CEEFC), University of Southern Queensland (USQ), Toowoomba, QLD

Talal Yusaf (water treatment using micro-organism; ultrasonic applications in water treatment), Executive Director, USQ International and Development, University of Southern Queensland (USQ), Toowoomba, QLD

AUSTRIA

Alfred Paul Blaschke (groundwater modeling and management; contaminant transport and behavior in the subsurface with a focus on microbiology and micropollutants; effects of infiltration of treated wastewater; integrated approaches: river – floodplain – infiltration – groundwater – groundwater use; climate change effects on water resources; thermal waters for water resource purposes), Deputy Head of the Interuniversity Cooperation Centre Water & Health (ICC); Centre for Water Resource Systems, Institute of Hydrology and Water Resource Management, Vienna University of Technology (TU Wien), Vienna

AZERBAIJAN

Farda Imanov (integrated water resource management, including assessment of environmental flows), Chairperson, UNESCO-IHP National Committee of Azerbaijan; Dean, Geography Department, Chair of Hydrometeorology, Baku State University (BSU), Baku

Rashail Ismayilov (hydrometeorology; hydrology; hydroecology; water resources management and water supply; ecology and environment; climate change), "Azersu" Joint Stock Company, Water Canal Scientific Research and Design Institute, Baku

BAHRAIN

Waleed Al-Zubari (water resources; environmental awareness; sustainable development; climate change), Coordinator of the UNU Water Virtual Learning Center for the Arab Region; Water Resources Management Division, Arabian Gulf University (AGU), Manama

BANGLADESH

Rafiqul Islam (natural resources; environment; climate change mitigation and management; energy technology), Dean, Faculty of Engineering & Technology, Director, Center for Climate Change Study and Resource Utilization, Chairman, Department of Nuclear Engineering, University of Dhaka (DU), Dhaka

BELARUS

Valiantsin Ramanouski (water resources; water and wastewater treatment; environmental engineering and science; sustainability; environmental awareness, impact and compliance), Department of Industrial Ecology, Belarusian State Technological University (BSTU), Central Research Institute for Complex Use of Water Resources, Minsk

BELGIUM

Michel Penninckx (depollution of agroindustrial effluents; bioremediation; valorization of effluents), Unit of Microbial Physiology and Ecology (UPEM), Free University of Brussels (ULB), Brussels

Enrico Ulisse Remigi (wastewater process modeling), DHI Group, Merelbeke

BELIZE

Ulric O'D Trotz (climate change impacts on small islands including water resources; adaptation across multi sectors and coordinating response to climate change impacts), Deputy Director, Caribbean Community Climate Change Centre, Belmopan

BOLIVIA

Carlos E. Román Calvimontes (hydrology; water and environmental management and modeling; wastewater use for irrigation), Center for Aerospace Survey and GIS Applications for Sustainable Development of Natural Resources (CLAS), University of San Simón (UMSS); Consultant, Cochabamba

BOSNIA AND HERZEGOVINA

Tarik Kupusović (water and environment management; hydraulics; resources efficiency; karst environment), Civil Engineering Faculty, University of Sarajevo (UNSA); Director, Hydro-Engineering Institute Sarajevo; President, Center for Environmentally Sustainable Development, Sarajevo

BOTSWANA

Berhanu F. Alemaw (sustainable water development, management and integrated modeling; surface and groundwater resources, use and management in agriculture; water supply; power and climate change mitigation; GIS and remote sensing), Hydraulic & Water Resources Engineering, Department of Geology, University of Botswana (UB), Gaborone

BRAZIL

Virginia S.T. Ciminelli (wastewater and tailings in mining and metallurgy: environmental impact assessment, effluent treatment; aqueous processing of materials), Department of Metallurgical and Materials Engineering & Director, National Institute of Science and Technology on Mineral Resources, Water and Biodiversity (INCT-Acqua), Federal University of Minas Gerais (UFMG), Belo Horizonte, MG

Luiz R.G. Guilherme (water, wastewater and soil in mining areas: contamination assessment, bioavailability, bioaccessibility and mitigation), Soil Science Department, Federal University of Lavras (UFLA), Lavras, MG

Erich Kellner (wastewater stabilization ponds; sustainability indicators to wastewater treatment systems), Department of Civil Engineering, Center of Exact Sciences and Technology, Federal University of São Carlos (UFSCar), São Carlos, SP

Eduardo Cleto Pires (aerobic and anaerobic wastewater treatment reactors), Department of Hydraulic and Sanitary Engineering, São Carlos School of Engineering, University of São Paulo (USP), São Paulo, SP

Jerusa Schneider (environmental sciences; phytoremediation; phytoextration; soil management, soil pollution, soil quality bioindicators, trace elements and nutrients from plants; influence of sanitary wastewater disinfection in functional diversity of soil microbial communities and vegetable production), Sanitation and Environment Department, School of Civil Engineering, Architecture and Urban Design, State University of Campinas (UNICAMP), Campinas, SP

BRUNEI

Jaya Narayan Sahu (water treatment using engineered low-cost natural materials, e.g. biogenic waste; sustainable wastewater and waste management; synthesis of nanomaterials for water and wastewater treatment; renewable energy; water-energy nexus; microwave energy application in chemical reaction engineering), Petroleum and Chemical Engineering

Program, Faculty of Engineering, Institute Technology Brunei (ITB), Our National Engineering and Technology University, Brunei Darussalam

BULGARIA

Roumen Arsov (technologies for wastewater treatment; sludge treatment and management; sewer and drainage networks and facilities), Bulgarian Water Association; Faculty of Hydraulic Engineering, Department of Water Supply, Sewerage, Water and Wastewater Treatment, University of Architecture, Civil Engineering and Geodesy (UACEG), Sofia

CAMEROON

Samuel N. Ayonghe (hydrogeology; environmental geology; geophysics), Department of Environmental Science, Faculty of Science, University of Buea (UB), Buea

Mathias Fru Fonteh (water resources management; water policy), Head, Department of Agricultural Engineering, Faculty of Agronomy and Agricultural Sciences, University of Dschang (UD), Dschang, West Region

CANADA

David Bethune (hydrogeology; water resources management; rural water supply and sanitation; hydrogeology/water resource management training programs in Latin America), founding member, Hydrogeologists Without Borders; International Centre, University of Calgary (U of C), Calgary, AB

Jeffrey J. McDonnell (isotope hydrology; ecohydrology; hillslope hydrology; forest hydrology), Global Institute for Water Security, University of Saskatchewan (U of S), Saskatoon, SK

Gunilla Öberg (sustainable sewage management in growing urban areas; water futures for sustainable cities; use of science in policy; chlorine biogeochemistry in soil), Institute for Resources, Environment and Sustainability (IRES), University of British Columbia (UBC), Vancouver, BC.

Rajinikanth Rajagopal (sustainable agro-food industrial wastewater treatment), Dairy and Swine Research and Development Centre, Agriculture and Agri-Food Canada, Quebec, QC

CHILE

Lorena Cornejo (distribution and dynamics of the chemical elements in water and soil; development, optimization and application of spectroscope techniques for environmental matrices; decontamination technologies and water disinfection), Laboratory of Environmental Research on Arid Zones (LIMZA, EUDIM), University of Tarapacá (UTA) & Environmental Resources Area, Research Center of the Man in the Desert (CIHDE), Arica

James McPhee (snow and mountain hydrology; hydrologic modeling; water resources engineering), Department of Civil Engineering, Faculty of Physical and Mathematical Sciences, University of Chile; Chilean focal point for water in the InterAmerican Network of Academies of Sciences (IANAS), Santiago de Chile

Bernabe Rivas (adsorption based water treatment), Faculty of Chemistry & Vice-Rector, University of Concepción (UdeC), Concepción

CHINA

Huaming Guo (water quality; hydrogeology; biogeochemistry; bioaccessibility and in-vivo bioavailability to trace contaminants; geochemistry; stable isotopes; environmental engineering; environmental chemistry), School of Water Resources and Environment, China University of Geosciences – Beijing (CUGB), Beijing

Hong-Ying Hu (wastewater reclamation and reuse; quality evaluation and risk management of reclaimed water; biological/ecological technologies for environmental pollution control and biomass/bio-energy production (micro-algae based wastewater treatment and biomass/bio-fuel production)), School of Environment, Tsinghua University (THU), Beijing

Kuan-Yeow Show (accelerated startup and operation of anaerobic reactors; microbial granulation in wastewater treatment; ultrasound applications in sludge and wastewater treatment; conversion of sludge and wastes into engineering materials), Department of Environmental Science and Engineering, Fudan University (FDU), Shanghai

Eddy Yong-Ping Zeng (inter-compartmental processes and fluxes of persistent organic pollutants (POPs); bioaccumulation and foodweb transfer of POPs; feasibility of using solid-phase microextraction-based technologies to measure freely dissolved concentrations of POPs in sediment porewater; human exposure and health risk), Dean, School of Environment, Jinan University (JNU), Guangzhou

COLOMBIA

Claudia Patricia Campuzano Ochoa (water and environment: research, planning, development, policy framework and strategic management; environmental integrated management of water resources and territorial dynamics; water footprint; inter-institutional coordination; education and environment; culture), Director of Water and Environment, and Coordinator of the Inter-Institutional Water Agreement at the Antioquia in the Science and Technology Center Corporation, Medellin, Antioquia

Gabriel Roldán Pérez (limnology and water resources), Colombian focal point for water in the InterAmerican Network of Academies of Sciences (IANAS); Research Group in Limnology and Water Resources, Catholic University of Oriente (UCO), Antioquia

COSTA RICA

Guillermo Alvarado Induni (geothermal fluids; water chemistry; balneology; interrelations between water chemistry and seismic/volcanic activity), Head, Seismology and Volcanology, Costa Rican Institute of Electricity (ICE), San Jose

CROATIA

Ognjen Bonacci (hydrology; karst hydrology; ecohydrology), Faculty of Civil Engineering, Architecture, and Geodesy, University of Split, Split

CYPRUS

Soteris Kalogirou (solar energy collectors; solar energy for seawater desalination; combination of CSP with desalination), Department of Mechanical Engineering and Materials Sciences and Engineering, Cyprus University of Technology (CUT), Limassol

CZECH REPUBLIC

Barbora Doušová (water and wastewater treatment; adsorption technologies; study, preparation, modification and testing of new sorbents to the verification of adsorption properties of natural matter – soils, sediments, dusts, etc.; geochemical processes on the solid-liquid interface), Department of Solid State Chemistry, University of Chemistry and Technology Prague (UCT Prague), Prague

Nada Rapantova (water resources and water quality; hydrogeology; groundwater pollution; groundwater modeling; mine water technology; uranium mining), Faculty of Mining and Geology, VSB – Technical University of Ostrava, Ostrava

Tomáš Vaněk (plant biotechnology; natural products chemistry; environment protection; phytotechnologies), Head, Laboratory of Plant Biotechnologies, Institute of Experimental Botany, Czech Academy of Sciences, Prague

DEMOCRATIC REPUBLIC OF THE CONGO

Dieudonné Musibono (ecotoxicology; ecosystems health; natural resources management; aquatic biodiversity; environmental & social impact assessment of projects; sustainable development; mining environment monitoring), Head, Department of Environmental Science and Engineering, Faculty of Science, University of Kinshasa; Former Programme Advisor and National Coordinator at UNEP, Kinshasa

DENMARK

Hans Brix (constructed wetlands for the treatment of polluted water; urban stormwater reuse; sustainable water/wastewater management in developing countries), Department of Bioscience, Aarhus University (AU), Aarhus

DOMINICAN REPUBLIC

Osiris de León (urban and industrial water pollution; water scarcity; groundwater exploration and exploitation; geophysical methods; dam sites exploration; environmental assessment for water pollution identification), Dominican focal point for water in the InterAmerican Network of Academies of Sciences (IANAS); Commission of Natural Sciences and Environment of the Science Academy, Academy of Sciences of the Dominican Republic, Santo Domingo

EGYPT

Fatma El-Gohary (water reuse), Water Pollution Research Department, National Research Centre, Dokki

Mohamed Fahmy Hussein (isotope hydrology and geochemistry applied in the holistic approach on surface and groundwater resources use; conservation and improvement of water quality; water management by integrating unsaturated and saturated zones and the potential promotion of water users implication in the control of water resources in the irrigated soils via water treatment and reuse), Soil and Water Department, Faculty of Agriculture, Cairo University (CU), Cairo

ESTONIA

Ülo Mander (landscape ecology (nutrient cycling at landscape and catchment levels) and ecological engineering (constructed wetlands and riparian buffer zones: design and performance), Department of Geography, Institute of Ecology and Earth Sciences, University of Tartu (UT), Tartu

ETHIOPIA

Tesfaye Tafesse (transboundary water issues, with emphasis on the Nile; natural resources management and institutions; rural development and agricultural problems in the Third World), College of Social Sciences, Addis Ababa University (AAU), Addis Ababa

Taffa Tulu (watershed hydrology; watershed management; water resources engineering; irrigation engineering; water harvesting), Center of Environment and Development, College of Development Studies, Addis Ababa University (AAU), Addis Ababa

FEDERATED STATES OF MICRONESIA

Leerenson Lee Airens (water supply for Small Islands Development States (SIDS)), GEF IWRM Focal Point; Manager, Water Works, Pohnpei Utilities Corporation (PUC), Pohnpei State

FIJI

Johann Poinapen (water and wastewater engineering and management; design and operation of water and wastewater treatment plants including membrane systems (MF & RO); brine treatment (thermal technologies); mine water treatment; water recycling), Acting Director, Institute of Applied Sciences, University of the South Pacific (USP), Suva

FINLAND

Riku Vahala (drinking water quality and treatment), Water and Environmental Engineering, Department of Civil and Environmental Engineering, School of Engineering, Aalto University, Aalto

FRANCE

Catherine Faur (water treatment involving fluid-solid interactions; engineering of polymer membranes by twin product – processes approaches), Department Engineering of Membrane Processes, University of Montpellier (UM), Montpellier

GEORGIA

Givi Gavardashvili (water management; erosion-debris flow processes; floods), Ts. Mirstkhulava Water Management Institute, Georgian Technical University (GTU), Tbilisi

GERMANY

Regina Maria de Oliveira Barros Nogueira (water and wastewater biology), Institute for Sanitary Engineering and Waste Management, Leibnitz University Hannover, Hannover

Jan Hoinkis (membrane technologies; membrane bioreactor technology; water and wastewater treatment; water reuse; sensor and control systems in water treatment), Institute of Applied Research, Karlsruhe University of Applied Sciences (HsKA), Karlsruhe

Heidrun Steinmetz (resource oriented sanitation (nutrient recovery, energy efficiency); biological and advanced wastewater treatment; water quality management), Chair of Sanitary Engineering and Water Recycling, University of Stuttgart, Stuttgart

GREECE

Maria Mimikou (hydrology; water resources management; hydro-energy engineering; climate change), School of Civil Engineering, National Technical University of Athens (NTUA), Athens

Anastasios Zouboulis (water and wastewater treatment; biotechnological applications), School of Chemistry, Aristotle University of Thessaloniki (AUTH), Thessaloniki

HAITI

Urbain Fifi (hydrogeology; environment engineering; groundwater quality and pollution; water resources management; hydrogeological modeling), President of IHP Haitian National Committee for UNESCO; Head of Research Master in "Ecotoxicology, Environment and Water Management", Faculty of Sciences, Engineering and Architecture, University Quisqueya, Haut de Turgeau, Port-au-Prince

HONDURAS

Sadia Iraisis Lanza (water resources and climate change; physical hydrogeology; hydrology; water quality), Physics Department, National Autonomous University of Honduras (UNAH), San Pedro Sula, Cortés

HONG KONG

Jiu Jimmy Jiao (hydrogeology; influence of groundwater and rainfall on landslides; impact of human activities on groundwater regimes; dewatering system design; contaminant fate and transport modeling and groundwater remediation design; global optimization approaches for parameter identification in flow and transport modeling; parameter sensitivity analysis and its influence on parameter estimation), Editor Hydrogeology Journal; Department of Earth Sciences, The University of Hong Kong (HKU), Hong Kong

HUNGARY

László Somlyódy (wastewater treatment; environmental engineering), past President of the International Water Association (IWA), Head, Department of Sanitary and Environmental Engineering, Faculty of Engineering, Budapest University of Technology and Economics (BME), Budapest

INDIA

Makarand M. Ghangrekar (wastewater treatment in microbial fuel cell and electricity generation), Department of Civil Engineering, Indian Institute of Technology – Kharagpur (IIT Kgp), Kharagpur, West Bengal

Arun Kumar (environmental management of water bodies), Alternate Hydro Energy Centre, Indian Institute of Technology – Roorkee (IITR), Roorkee, Uttarakhand

Rakesh Kumar (urban hydrology; hydrological modeling; watershed management; drought mitigation and management; flood estimation, routing, management and socio-economic aspects; impact of climate change on water resources), Head, Surface Water Hydrology Division, National Institute of Hydrology (NIH), Roorkee, Uttarakhand

Abhijit Mukherjee (physical, chemical and isotope hydrogeology; modeling of groundwater flow and solute transport; hydrostratigraphy; contaminant fate and transport; surface water-seawater-groundwater interactions; effect of climate change on water resources; mine-site hydrology; environmental geochemistry), Department of Geology and Geophysics, Indian Institute of Technology – Kharagpur (IIT Kgp), Kharagpur, West Bengal

INDONESIA

Budi Santoso Wignyosukarto (water resources; low land hydraulics, mathematical modeling), Department of Civil & Environmental Engineering, Faculty of Engineering, Gagjah Mada University (UGM), Yogyakarta

IRAN

Ahmad Abrishamchi (water resources and environmental systems: analysis and management), Chairholder, UNESCO Chair in Water and Environment Management for Sustainable Cities; Department of Civil Engineering, Sharif University of Technology (SUT), Tehran

ISRAEL

Ofer Dahan (vadose zone and groundwater hydrology; quantitative assessment of water infiltration and groundwater recharge; water flow and contaminant transport through the vadose zone; arid land hydrology; monitoring technologies for the deep vadose zone), Department of Hydrology & Microbiology, Zuckerberg Institute for Water Research, Blaustein Institute for Desert Research, Ben Gurion University of the Negev (BGU), Sde Boker Campus, Ben Gurion

Michael Zilberbrand (groundwater resources; hydrogeochemical processes and hydrogeological and hydrogeochemical modeling in aquifers and in the unsaturated zone), Israeli Water Authority, Hydrological Service, Jerusalem

ITALY

Alessandra Criscuoli (membrane science and technology; membrane distillation and membrane contactors; integrated membrane processes; water and wastewater treatment; desalination of brackish water and seawater), Institute on Membrane Technology, ITM-CNR, Rende (CS)

Enrico Drioli (membrane science and engineering; membrane preparation and transport phenomena in membranes; desalination of brackish and saline water; integrated membrane processes; membrane distillation and membrane contactors; catalytic membrane and catalytic membrane reactors; salinity gradient energy fuel cells), Institute on Membrane Technology, ITM-CNR, Rende (CS)

Alberto Figoli (membrane science and technology; membrane preparation and characterization; transport phenomena in membranes; pervaporation; water and wastewater treatment; desalination of brackish and saline water), Institute on Membrane Technology, ITM-CNR, Rende (CS)

Marco Petitta (groundwater pollution, management, and protection), President IAH Chapter Italy; Department of Earth Sciences, Sapienza University of Rome, Rome

Ludovico Spinosa (sludge management), (retired) National Research Council (CNR); Consultant at Governmental Commissariat Environmental Emergencies in Region Puglia; Convenor at ISO/TC275/WG6 (Thickening and Dewatering) and CEN/TC308/WG1 (Process Control Methods) on sludge standardization

JAMAICA

Arpita Mandal (hydrology; hydrogeology; water resources and impacts of climate change; water supply; climate variability; flood risk and control; hydrochemistry of groundwater; saline water intrusion), Department of Geography and Geology, University of the West Indies (UWI), Mona Campus, Mona, Kingston

JAPAN

Hiroaki Furumai (build-up and wash-off of micropollutants in urban areas; characterization of DOM/NOM in lakes and reservoirs for drinking water sources; fate and behavior of DOM in flocculation and advanced oxidation processes; biological nutrient removal from wastewater; modeling activated sludge in aerobic/anaerobic SBR; characterization of domestic sewage from the viewpoint of nutrient removal), Board of Directors, IWA; Department of Urban Engineering, The University of Tokyo (Todai), Tokyo

Makoto Nishigaki (modeling groundwater and multiphase flow and solute transport in porous media; modeling seepage in the saturated-unsaturated zone; development of methods of measuring hydraulic properties in rock mass), Department of Environmental and Civil Design, Faculty of Environmental Science and Technology, Okayama University, Okayama

Taikan Oki (global water balance and world water resources; climatic variation and the Asian monsoon; land-atmosphere interaction and its modeling; remote sensing in hydrology; temporal and spatial distribution of rainfall), Institute of Industrial Science, The University of Tokyo, Komaba, Tokyo

Yuichi Onda (hillslope hydrology; hydro-geomorphology; radionuclide transfer; forest hydrology), Center for Research in Isotopes and Environmental Dynamics, University of Tsukuba, Tsukuba, Ibaraki

Kaoru Takara (innovative technologies for predicting floods; global environmental changes; risk and emergency management; interactions between social changes and hydrological cycle/water-related disasters; disaster mitigation strategy; policy development; integrated numerical modeling for lakes and surrounding catchments), Director, Disaster Prevention Research Institute, Kyoto University (Kyodai), Kyoto

JORDAN

Fawzi A. Banat (desalination), Department of Chemical Engineering, Jordan University of Science and Technology (JUST), Irbid

Samer Talozi (irrigation and water resources engineering, planning and policy), Civil Engineering Department, Jordan University of Science and Technology (JUST), Irbid

KENYA

Daniel Olago (environmental geology; surface and sub-surface water chemistry and dynamics; water-energy and related nexuses; human impact on the environment, global change processes, vulnerability and adaptation to climate change: past and present; environmental policies, laws and regulations and capacity development in global environmental change), Chairman, Network of African Science Academies (NASAC) Water Program; Member, International Lake Environment Committee; Member and focal point for water, Kenya National Academy of Sciences (KNAS); Institute for Climate Change and Adaptation (ICCA) & Department of Geology, University of Nairobi, Nairobi

Mwakio Tole (water and geothermal energy resources; waste disposal; environmental impact assessment), School of Environmental and Earth Sciences, Department of Environmental Sciences, Pwani University, Kilifi

KOREA

Jaeweon Cho (water reuse; membrane filtration; ecological engineering (treatment wetland); desalination), School of Urban and Environmental Engineering, Ulsan Institute of Science and Technology (UNIST), Ulsan

KYRGYZSTAN

Bolot Moldobekov (hydrogeology; engineering geology; geographic information systems – GIS; geoinformatics; interdisciplinary geosciences; natural hazards), Co-Director, Central-Asian Institute for Applied Geosciences (CAIAG), Bishkek

LATVIA

Māris Kļaviņš (aquatic chemistry; geochemical analysis; environmental pollution and its chemical analysis; environmental education, including also political and social sciences), Head, Department of Environmental Science, University of Latvia (LU), Riga

LITHUANIA

Robert Mokrik (groundwater resources, flow and transport modeling; hydrogeochemistry and groundwater isotopes; palaeohydrogeology), Department of Hydrogeology and Engineering Geology, Faculty of Natural Sciences, Vilnius University, Vilnius

LUXEMBOURG

Joachim Hansen (wastewater treatment; micropollutants; wastewater reuse; water-energy nexus), Engineering Science – Hydraulic Engineering, Faculty of Science, Technology and Communication, University of Luxembourg – Campus Kirchberg, Luxembourg

MADAGASCAR

Désiré Rakotondravaly (hydrology; hydrogeology; hydraulics; geology; rural water supply; vulnerability mapping; water and sanitation; GIS; project management; capacity building; community development; conservation; development cooperation), Ministry of Mines, Antananarivo

MALAWI

Victor Chipofya (urban water utility operation and management; groundwater development, monitoring and management; groundwater quality; rural water supply; water and sanitation in peri-urban and rural areas; water reuse; hygiene promotion), Executive Director, Institute of Water and Environmental Sanitation (IWES); National Coordinator of the Malawi Water Partnership (MWP); Steering Committee Member: Water Supply and Sanitation Collaborative Council (WSSCC) for Eastern and Southern Africa, Blantyre

MALAYSIA

Mohamed Kheireddine Aroua (separation processes; water and wastewater treatment), Director, Centre for Separation Science & Technology (CSST), Department of Chemical Engineering, Faculty of Engineering, University of Malaya (UM), Kuala Lumpur

Hamidi Abdul Aziz (water supply engineering; wastewater engineering; solid waste management), School of Civil Engineering, University of Science Malaysia (USM), Engineering Campus, Nibong Tebal, Penang

Ali Hashim (separation processes – flotation; liquid-liquid extraction; water and wastewater treatment; ionic liquids – synthesis and applications), Department of Chemical Engineering, Faculty of Engineering, University of Malaya (UM), Kuala Lumpur

Ahmad Fauzi Ismail (development of membrane technology for reverse osmosis, nanofiltration, ultrafiltration and membrane contactor), Deputy Vice Chancellor (Research & Innovation) & Founder and Director, Advanced Membrane Technology Research Center (AMTEC), University of Technology Malaysia (UTM), Johor Bahru, Kuala Lumpur

Hilmi Mukhtar (membrane development; membrane modeling; membrane applications including wastewater treatment engineering and natural gas separation), Department of Chemical Engineering, Faculty of Engineering, Petronas University of Technology (UTP), Bandar Seri Iskandar, Perak

Mohd Razman Bin Salim (water and wastewater treatment), Deputy Director, Centre for Environmental Sustainability and Water Security (IPASA), Faculty of Civil Engineering, University of Technology Malaysia (UTM), Johor Bahru, Johor

Saim Suratman (hydrogeology; groundwater management), Deputy Director General, National Hydraulics Research Institute of Malaysia (NAHRIM), Seri Kembangan Selangor Darul Ehsan, Malaysia

Wan Azlina Wan Ab Karim Ghani (chemical and environmental engineering; biochar and composites for water, wastewater and soil treatment; biomass conversion; biomass energy), Research Coordinator, Department of Chemical & Environmental Engineering, Faculty of Engineering, Putra University Malaysia (UPM), Serdang

MALTA

Kevin Gatt (governance, policy and planning issues related to water resources; waste management and sustainable development), Faculty for the Built Environment, University of Malta (UoM), Tal-Qroqq, Msida

MAURITIUS

Arvinda Kumar Ragen (wastewater engineering; constructed wetlands for household greywater; water pollution control in sugar factories; environmental impact assessment), Department of Chemical & Environmental Engineering, Faculty of Engineering, University of Mauritius (UoM), Le Reduit, Moka.

MEXICO

Ma. Teresa Alarcón Herrera (water resources; water treatment using artificial wetlands), Director, Durango Unit of the Advanced Materials Research Center (CIMAV), Durango, Dgo.

Maria Aurora Armienta (hydrogeology; trace element contaminants; water treatment using geological materials), Institute of Geophysics, National Autonomous University of Mexico (UNAM), Ciudad Universitaria, Mexico City, D.F.

Sofia Garrido Hoyos (drinking water; collection and treatment of rainwater; biological wastewater treatment; treatment and/or utilization of sludge and biosolids), Mexican Institute of Water Technology (IMTA), Jiutepec, Mor.

Luz Olivia Leal Quezada (environmental engineering; environmental chemistry; automation of chemical analysis techniques for environmental monitoring, particularly for the determination and speciation of trace elements; techniques for determining water quality and chemical aspects of their treatment), Advanced Materials Research Center (CIMAV), Environment and Energy Department, Chihuahua, Chih.

MOROCCO

Lhoussaine Bouchaou (hydrology; water quality; aquatic ecosystems; environmental impact assessment; climatology; climate change), President IAH Chapter Morocco; Applied Geology and Geo-Environment Laboratory, Faculty of Sciences, University Ibn Zohr (UIZ), Agadir

MOZAMBIQUE

Catine Chimene (municipal water and infrastructure; water supply engineering; agricultural water; rural development), Higher School of Rural Development (ESUDER), Eduardo Mondlane University (UEM), Inhambane, Vilankulo

MYANMAR

Khin-Ni-Ni Thein (hydroinformatics, integrated water resources management, river basin management, coastal-zone management, sustainable hydropower assessment, disaster risk reduction, climate change; sustainability; capacity building; community development; water and environmental policy; public policy analysis; green economy and green growth), Secretary, Advisory Group, Member, National Water Resources Committee; Advisory Group Member, National Disaster Management Committee; Founder and President, Water, Research and Training Centre (WRTC); Visiting Senior Professor, Yangon Technological University (YTU), Yangon, Myanmar; Regional Water Expert for Green Growth, UNESCAP

NAMIBIA

Benjamin Mapani (groundwater recharge and vulnerability mapping; groundwater development, management, monitoring and modeling; environmental hydrogeology; climate change), Board of Trustees, WaterNet; Department of Geology, University of Namibia (UNAM), Windhoek

NEPAL

Bandana K. Pradhan (environment and public health), Department of Community Medicine and Public Health, Institute of Medicine, Tribhuvan University (TU), Maharajgunj

NEW ZEALAND

David Hamilton (modeling of water quality in lakes and reservoirs; sediment-water interactions in lakes; boom-forming algae, particularly cyanobacteria; ice cover in lakes), Environmental Research Institute (ERI), University of Waikato, Waikato

NICARAGUA

Andrew Longley (hydrogeology; groundwater engineering; catchment studies and groundwater modeling; international development: projects in the water, geothermal, agriculture, environment and health sectors; rural water supply; arsenic contamination: mapping, hydrogeology, epidemiology; bridging the gap between academia, industry, public and charity sectors), Director, Nuevas Esperanzas UK, León

Katherine Vammen (aquatic microbiology; climate change and water resources; water supply and sanitation for the poor; urban waters), Co-Chair of the Water Programme of the Interamerican Network of the Academies of Science; Nicaraguan focal point for water programme in the InterAmerican Network of Academies of Sciences (IANAS); Central American University, Managua

NIGERIA

Peter Cookey (sustainable water and wastewater management in developing countries), Rivers State College of Health Science and Technology, Port Harcourt, Nigeria and Earthwatch Research Institute (EWRI), Port Harcourt

NORWAY

Torleiv Bilstad (water, oil and gas separation; environmental science and engineering), Former President of EWA-Norway; Department of Mathematics and Natural Sciences, University of Stavanger (UiS), Stavanger

Hallvard Ødegaard (water and wastewater treatment; innovative solutions for integrated approaches to urban water management), Department of Hydraulic and Environmental Engineering, Norwegian University of Science and Technology (NTNU), Trondheim

OMAN

Mohammed Zahir Al-Abri (thermal desalination; water and wastewater treatment; nanotechnology), Petroleum and Chemical Engineering Department, Sultan Qaboos University (SQU), Al Khoudh, Muscat

PAKISTAN

Ghani Akbar (agricultural engineering; integrated water management; soil and water conservation and climate-smart agricultural practices), Program Leader, Integrated Watershed Management Program (IWMP), Climate Change, Alternate Energy and Water Resources Institute (CAEWRI), National Agricultural Research Centre (NARC), Chak Shahzad, Islamabad

PALESTINIAN AUTONOMOUS AREAS

Marwan Haddad (interdisciplinary approaches to water resources and quality management; renewable energy; recycling), Director, Water and Environmental Studies Institute, An Najah National University, Nabus

PANAMA

José R. Fábrega (sustainable water and wastewater management; environmental fate of chemicals in water and soil systems), Panamanian focal point for water in the InterAmerican Network of Academies of Sciences (IANAS); Hydraulic and

Hydrotechnical Research Center (CIHH), Technological University of Panama (UTP), Panama City

PARAGUAY

Alicia Eisenkölbl (environmental management; environmental impact assessment; trans-boundary aquifers; rural development), Faculty of Agricultural Sciences Hohenau, Catholic University Our Lady of the Assumption (UCA), Campus Itapúa, Encarnación

PERU

Nicole Bernex Weiss de Falen (integrated water resources management; human sustainable development; climate change adaptation; integrated ecosystemic services, water management and risks (droughts and floods) with land planning at a water basin, regional and national level), Peruvian focal point for water in the InterAmerican Network of Academies of Sciences (IANAS); member of the technical Committee of Global Water Partnership GWP; LAC Chair in the CST of the UNCCD; Center of Research in Applied Geography (CIGA), Pontifical Catholic University of Peru (PUCP), Lima

PHILIPPINES

Victor Ella (surface and groundwater hydrology; irrigation and drainage engineering; water quality; simulation modeling; wastewater engineering; contaminant transport in soils; geostatistics; hydraulic engineering), Land and Water Resources Division, Institute of Agricultural Engineering, College of Engineering and Agro-Industrial Technology, University of the Philippines Los Baños (UPLB), College, Laguna

POLAND

Marek Bryjak (adsorption based water treatment), Department Polymer & Carbon Materials, Wrocław University of Technology, Wrocław

Wieslaw Bujakowski (geothermics), Mineral and Energy Economy Research Institute, Polish Academy of Sciences (PAS), Kraków

Jacek Makinia (wastewater treatment; nutrient removal and recovery from wastewater), Faculty of Hydro and Environmental Engineering, Vice-Rector for Cooperation and Innovation, Gdańsk University of Technology (GUT), Gdańsk

Barbara Tomaszewska (monitoring of the aquatic environments; geothermics; scaling of geothermal systems; membrane technologies for geothermal water treatment for water resource purposes), AGH University of Science and Technology; Mineral and Energy Economy Research Institute, Polish Academy of Sciences (PAS MEER), Kraków

PORTUGAL

Maria do Céu Almeida (sewer processes and networks), National Laboratory of Civil Engineering (LNEC), Lisbon

Helena Marecos (water reuse), Civil Engineering Department, Lisbon Engineering Superior Institute (ISEL), Lisbon

Helena Ramos (water-energy nexus; energy efficiency and renewable energies; hydraulics; hydrotransients; hydropower; pumping systems; leakage control; water supply; water vulnerability), Department of Civil Engineering, University of Lisbon (ULisboa), Lisbon

QATAR

Farid Benyahia (immobilized nitrifiers in wastewater treatment; membrane distillation desalination; water quality and energy efficiency analysis; airlift bioreactors; low-grade heat in membrane distillation for freshwater production; bioremediation of oil spills; development, design and evaluation of advanced refinery wastewater treatment processes), College of Engineering, Department of Chemical Engineering, Qatar University (QU), Doha

Patrick Linke (design, engineering and optimization of efficient processes, integrated systems and associated infrastructures; efficient utilization of natural resources (energy, water and raw materials); water-energy-food nexus), Chair, Chemical Engineering Program, Texas A&M University at Qatar (TAMUQ), Managing Director of the Qatar Sustainable Water and Energy Utilization Initiative (QWE) at TAMUQ, Qatar Environment and Energy Research Institute (QEERI), Doha

REPUBLIC OF GUINEA

Hafiziou Barry (integrated water resources management), Polytechnic Institute, University Gamal Abdel Nasser, Conakry

ROMANIA

Anton Anton (pumping stations; municipal water networks), Hydraulics and Environmental Protection Department, Technical University of Civil Engineering (UTCB), Bucharest

RUSSIAN FEDERATION

Sergey Pozdniakov (water resources; water quality; hydrogeology; contaminant transport; geostatistics; water balance; climate change), Faculty of Geology, Moscow State University (MSU), Moscow

RWANDA

Omar Munyaneza (hydrology; climate change and water resources management), College of Science and Technology, Department of Civil Engineering, University of Rwanda (UR), Kigali

SAUDI ARABIA

Noreddine Ghaffour (renewable energy for desalination and water treatment), Water Desalination and Reuse Research Center, King Abdullah University of Science and Technology (KAUST), Thuwal

Mattheus Goosen (renewable energy for desalination and water treatment; membranes), Office of Research and Graduate Studies, Alfaisal University, Riyadh

SENEGAL

Alioune Kane (water quality; hydraulics; water-poverty relationships; climate variability and water availability), Director of the Master Programme GIDEL (Integrated Management and Sustainable Development of Coastal West Africa); Coordinator of WANWATCE (Centres Network of Excellence for Science and Water Techniques NEPAD), Department of Geography, Cheikh Anta Diop University (UCAD), Dakar

SERBIA

Petar Milanović (karst hydrogeology; theory and engineering practice in karst), President IAH Chapter Serbia and Montenegro, Belgrade

SINGAPORE

Vladan Babovic (hydroinformatics; data assimilation; data mining), Department of Civil and Environmental Engineering, National University of Singapore (NUS), Singapore

Jiangyong Hu (water treatment technology; water quality; water reuse; health impacts), Department of Civil and Environmental Engineering & Co-Director, Centre for Water Research, National University of Singapore (NUS), Singapore

SLOVAKIA

Ján Derco (environmental engineering; nutrients removal; ozone-based oxidation processes; water resources protection; water and wastewater technology), Institute of Chemical and Environmental Engineering, Faculty of Chemical and Food Technology, Slovak University of Technology (SUT), Bratislava

SLOVENIA

Boris Kompare (wastewater treatment; modeling), Past President EWA-Slovenia; Faculty of Civil Engineering and Geodesy, University of Ljubljana (UL), Ljubljana

SOMALIA

Abdullahi Mohumed Abdinasir (water resources management; groundwater governance; water supply), Ministry of Water, Petroleum, Energy and Mineral Resources, Mogadishu

SOUTH AFRICA

Tamiru A. Abiye (community water supply problems; water quality assessment and monitoring; hydrochemical modeling; groundwater flow and solute transport; trace metals in groundwater; surface and groundwater interactions; climate change impact on groundwater; spatial and temporal variability of groundwater recharge), School of Geosciences, Faculty of Science (East Campus), University of the Witwatersrand (Wits University), Johannesburg

Hamanth C. Kasan (sustainable water and wastewater management in developing countries), General Manager, Scientific Services Division, Rand Water; President, African Water Association (AfWA), Johannesburg

Sabelo Mhlanga (water-energy nexus; nano-structured materials for water purification and recovery; energy-efficient and antifouling membrane filtration technologies for water treatment; community involvement in water related problems in rural communities; green chemistry), Deputy Director, Nanotechnology and Water Sustainability (NanoWS) Research Unit, College of Science Engineering and Technology, University of South Africa (Unisa), Johannesburg

Anthony Turton (water-energy-food nexus; hydropolitical risk model; mine water management; mine closure planning and strategies; groundwater governance; wastewater reuse), Director, Environmental Engineering Institute of Africa; Centre for Environmental Management, University of Free State (UFS), Bloemfontein; professor at UNESCO Chair in Groundwater, Department of Earth Sciences, University of Western Cape (UWC)

SPAIN

José Ignacio Calvo (membrane technologies; modifications of structure and surface properties of membranes to increase selectivity), School of Agriculture, Food Technology and Forestry, ETSIIAA, University of Valladolid (UVa), Palencia

Jesús Colprim (small water supply and wastewater systems), Laboratory of Chemical and Environmental Engineering (LEQUIA), Institute of Environment, University of Girona (UdG), Girona

Elena Giménez-Forcada (hydrogeology; hydrogeochemistry; water quality; groundwater contamination; trace elements), Geological Survey of Spain (IGME), Salamanca

J. Jaime Gómez-Hernández (stochastic hydrogeology; geostatistics; inverse modeling; nuclear waste disposal), Head of the Group of Hydrogeology, Research Institute of Water and Environmental Engineering, UPV, Valencia

Aurora Seco Torrecillas (nutrient removal and recovery from wastewater; anaerobic membrane bioreactor for wastewater treatment (WWT); microalgae cultivation for WWT), Chemical Engineering Department, University of Valencia (UV), Valencia

Guillermo Zaragoza (solar energy for desalination; thermal and membrane technologies for water treatment), Solar Platform of Almería (PSA-CIEMAT), Almería

SRI LANKA

Nadeeka S. Miguntanna (urban stormwater quality and monitoring; surrogate water quality parameters; urban water pollution; rainwater harvesting), Environmental Engineering Laboratory, Department of Civil and Environmental Engineering, Faculty of Engineering, University of Ruhuna (UOR), Hapugala, Galle

Meththika Suharshini Vithanage (water quality; water chemistry; impact of tsunamis on aquifers; groundwater modeling; soil and water monitoring for pollution; mechanistic modeling of biochar and nano materials for water and soil remediation (landfill leachates, nutrients and toxic metal(loid)s)), Group Leader – Chemical and Environmental Systems Modeling Research Group, National Institute of Fundamental Studies (NIFS), Kandy

SUDAN (REPUBLIC OF)

Abdin Mohamed Ali Salih (environmental sciences with emphasis on water resources management in arid and semi-arid zones), Board Member at UNESCO-IHE; Civil Engineering Department, Faculty of Engineering, The University of Khartoum (UofK), Khartoum

SURINAME

Sieuwnath Naipal (hydrology; climate change impact on climate variability; marine and coastal engineering), Anton De Kom University of Suriname (AdeKUS), Tammenga

SWAZILAND

Absalom M. Manyatsi (land and water resources management; environmental impact assessment; remote sensing; GIS and spatial climate change impacts; climate change adaptation and mitigation; climate smart agriculture; climate change and climate variability impacts on water resources and agriculture), Head of Department, Agricultural and Biosystems Engineering Department, University of Swaziland (UNISWA), Luyengo

SWEDEN

Prosun Bhattacharya (groundwater resources; hydrogeochemistry; arsenic), Coordinator, KTH-International Groundwater Arsenic Research Group, Department of Sustainable Development, Environmental Science and Engineering, Royal Institute of Technology (KTH), Stockholm

Joydeep Dutta (application of nanotechnology for water treatment; water-environment-energy nexus; photocatalytic materials, electrocatalysis and capacitive desalination of water, plasmon resonance sensors for heavy metal ion detection), Chair, Functional Materials Division, Materials- and Nano Physics Department, ICT School, Kista, Stockholm

Gunnar Jacks (hydrology; hydrogeology; hydrochemistry; groundwater chemistry; groundwater arsenic and fluoride; acidification of soil and groundwater; artificial groundwater recharge; water supply and sanitation in suburban areas), (retired) Department of Sustainable Development, Environmental Science and Engineering, Royal Institute of Technology (KTH), Stockholm

Erik Kärrman (sustainable wastewater management; decision support and multi-criteria analysis of municipal and local water and wastewater systems), Director of Research and Development, Urban Water & Royal Institute of Technology (KTH), Stockholm

Andrew Martin (membrane distillation for desalination and water purification; biomass and municipal solid waste; polygeneration), Department of Energy Technology, Royal Institute of Technology (KTH), Stockholm

Aapo Sääsk (development of polygeneration and zero liquid discharge), CEO, Scarab Development AB, Stockholm

Nury Simfors (geothermal exploration, geogenic trace contaminants, GIS and hazard mitigation), MVI/MTA, Swedish National Defence University (FHS), Stockholm

SWITZERLAND

Annette Johnson (geochemistry of inorganic contaminants in surface- and groundwater; chemistry of the mineral-water interface; geogenic contamination of groundwater and drinking water; waste management), Leader, Department of Water Resources and Drinking Water, Swiss Federal Institute of Aquatic Science and Technology (Eawag), Düebendorf

Eberhard Morgenroth (biological wastewater treatment using activated sludge, granular biomass, or biofilm based systems; biological drinking water treatment; mixed culture environmental biotechnology for bioenergy production; mathematical modeling; sustainable development of urban water management), Chair of Process Engineering in Urban Water Management, ETH Zurich & Swiss Federal Institute of Aquatic Science and Technology (Eawag), Dübendorf

Thomas Wintgens (water reuse), Institute for Ecopreneurship, School for Life Sciences, University of Applied Sciences and Arts Northwestern Switzerland (FHNW), Muttenz

TAIWAN

How-Ran Guo (epidemiology; environmental health and medicine; health impacts of water contaminants; cancers and chronic diseases in endemic areas of arsenic intoxication; epidemiologic characteristics of diseases associated with arsenic), Department of Environmental and Occupational Health, College of Medicine, National Cheng-Kung University (NCKU), Tainan

Tsair-Fuh Lin (analysis and monitoring of algae, cyanobacteria and metabolites; water treatment; site remediation; adsorption technology), Director, Global Water Quality Research Center, Department of Environmental Engineering, National Cheng-Kung University (NCKU), Tainan

Jyoti Prakash Maity (water and wastewater treatment with microbes and algae), Department of Earth and Environmental Sciences, National Chung Cheng University (CCU), Ming-Shung, Chiayi County

TANZANIA

Jamidu H.Y. Katima (sustainable wastewater management), College of Engineering and Technology, University of Dar es Salaam (UDSM), Dar es Salaam

THAILAND

Nathaporn Areerachakul (rainwater utilization; photocatalytic hybrid systems), Graduate School, Suan Sunandha Rajhabhat University (SSRU), Dusit, Bangkok

Mukand S. Babel (hydrologic and water resources modeling as applied to integrated water resources management; watershed modeling and management; water resources allocation and management; water resources and socio-economic development; water supply system and management; climate change impact and adaptation in the water sector), Coordinator, Water Engineering and Management (WEM), Director, COE for Sustainable Development in the Context of Climate Change (SDCC), Asian Institute of Technology (AIT), Pathumthani

Thammarat Koottatep (sustainable wastewater management in developing countries; decentralized waste and wastewater treatment systems; eco-engineering technology for wastewater treatment and management; environmental health and sanitation), Environmental Engineering and Management, School of Environment, Resources and Development, Asian Institute of Technology (AIT), Khlong Luang, Pathum Thani

Chongrak Polprasert (sustainable wastewater management technology; hazardous waste minimization and control; global warming and climate change mitigation and adaptation), Department of Civil Engineering, Faculty of Engineering, Thammasat University (TU), Bangkok

Wipada Sanongraj (water and air pollution control and modeling), Chemical Engineering Department, Ubon Ratchathani University (UBU), Ubon Ratchathani

Sangam Shrestha (hydrology and water quality model development; water and climate change impacts and adaptation; water-energy-climate change nexus; impacts of climate change on: agriculture and food security, urbanization, land use and forestry, water supply and sanitation, infrastructure, and energy security), Water Engineering and Management, School of Engineering and Technology, Asian Institute of Technology (AIT), Pathumthani

THE NETHERLANDS

Arslan Ahmad (removal of arsenic, chromium, fluoride, iron, manganese, pharmaceutical residues, emerging compounds from drinking water, and industrial water treatment for recovery of salts through eutectic freeze crystallization process), KWR Watercycle Research Institute, Nieuwegein

Tony Franken (membrane technologies; nanofiltration; reverse osmosis; membrane distillation), Director, Membrane Application Centre Twente (MACT bv), Enschede

Antoine J.B. Kemperman (membrane technology), Faculty of Science and Technology, University of Twente (UT), Enschede

TONGA

Taaniela Kula (water supply for Small Islands Development States (SIDS)), GEF- IWRM Focal Point; Deputy Secretary for Natural Resources Division, Ministry of Lands Survey and Natural Resources, Nuku'alofa

TRINIDAD AND TOBAGO

John Agard (wetland restoration; tropical small island ecology; socio-economic climate change mitigation and adaptation scenarios; ecosystem services; marine benthic ecology; pollution control), Tropical Island Ecology, Faculty of Science and Technology, The University of the West Indies (UWI), St. Augustine Campus

Everson Peters (water resources; water management; rainwater harvesting in Small Islands Developing States (SIDS); wastewater reuse; willingness to pay), The Faculty of Engineering, The University of The West Indies (UWI), St. Augustine Campus

TURKEY

Ibrahim Gurer (surface and groundwater engineering; snow hydrology; flood hydrology; ecohydrology; transboundary waters; climatic trend analysis; water politics), Department of Civil Engineering, Faculty of Engineering, Gazi University, Ankara

Nalan Kabay (water and wastewater treatment by ion exchange and membrane processes (NF, RO, ED, EDI, MBR); water reuse; desalination; boron separation), Chemical Engineering Department, Engineering Faculty, Ege University, Bornova, Izmir

UGANDA

Albert Rugumayo (hydrology; water resources engineering and management; energy management, policy and planning project design; renewable energy), Faculty of Engineering, Ndejje University and School of Engineering, CEDAT, Makerere University – Kampala (MUK), Kampala

UK

Rafid Alkhaddar (wastewater treatment; water reuse and conservation), Head, Department of Civil Engineering, Liverpool John Moores University (LJMU), Liverpool

J.A. [Tony] Allan (water resources and the political economy of water policy and its reform; water resources in food supply chains; hydro-politics), Department of Geography, King's College & SOAS Food Studies Centre, University of London, London

Vinay Chand (rural development; renewable energy cogeneration water purification; water reuse and recycling in the semiconductor industry; remediation of arsenic-contaminated well waters; developing polygeneration as a way of overcoming both technical and economic impediments to the food-energy-water cycle; low grade waste heat for membrane distillation), CEO of Xzero AB, Chairman of HVR Water Purification AB, Stockholm, Consultant Economist, London

Ian Griffiths (mathematical modeling of strategies for water purification, including membrane filtration and contaminant removal via magnetic separation), Mathematical Institute, University of Oxford, Oxford

Nidal Hilal (water treatment; desalination; membrane separation; engineering applications of atomic force microscopy; fluidization engineering), Director, Centre for Water Advanced Technologies and Environmental Research (CWATER), College of Engineering, Swansea University, Swansea

Chuxia Lin (acid mine drainage; water-soil-plant interaction), College of Science & Technology, University of Salford, Manchester

Victor Starov (influence of surface forces on membrane separation: nano-filtration, ultra- and microfiltration (fouling and/or gel layers formation)), Department of Chemical Engineering, Loughborough University, Loughborough

James Wilsdon (social science of the 'nexus': joined-up approaches to food, energy, water and environment; politics of scientific advice; interdisciplinarity, particularly between natural and social sciences; governance of new and emerging technologies; science and innovation policy; public engagement in science, technology and research), Science Policy Research Unit, University of Sussex, Brighton

UKRAINE

Valeriy Orlov (technology for potable water; water supply and sewerage; intakes; drilling; environmental aspects of water supply; system analysis), Academician of the International Academy of Ecology and Life Safety, Academician of the Academy of Building of Ukraine;

Academician of the Academy of Higher Education of Ukraine; National University of Water Management and Natural Resources (NUWMNRU), Rivne

UNITED ARAB EMIRATES

Fares M. Howari (water and soil management; environmental quality; fate and transport of contaminants; natural resources assessment and development; heavy metals; uranium geology; salinity; geochemistry; hydrology; remote sensing), Department of Natural Science and Public Health, Zayed University (ZU), Abu Dhabi

URUGUAY

Maria Paula Collazo (water resources management; water quality; hydrogeology; hydrogeochemistry; environmental awareness; sustainable development; groundwater remediation), Regional Centre for Groundwater Management for Latin America and the Caribbean (CeReGAS), MVOTMA/UNESCO; Faculty of Sciences, University of the Republic (UdelaR), Montevideo

USA

David Austin (emerging sustainable wastewater treatment technologies: design and operation; low-energy and positive-energy yield wastewater treatment; tidal-flow wetlands; improving raw water quality in reservoirs for drinking water plants), President (2015–2016), American Ecological Engineering Association; David Austin, Global Technology Lead Natural Treatment Systems; CH2M, Mendota Heights, MN

Leslie Behrends (decentralized wastewater technology; tidal-flow reciprocating wetlands), President, Tidal-flow Reciprocating Wetlands LLC; Consultant/Partner for ReCip systems, Florence, AL

Harry R. Beller (environmental engineering and microbiology; renewable fuels and chemicals; design and engineering of novel biofuel pathways in bacteria; biological treatment of contaminated groundwater; physiology and biochemistry of anaerobic bacteria; biodegradation and biotransformation of organic and inorganic contaminants (environmental biogeochemistry); development of mass spectrometric and biomolecular techniques to document in situ metabolism; hydraulic fracturing and water quality; water-energy nexus), Director of Biofuels Pathways in DOE's Joint BioEnergy Institute (JBEI); Senior Scientist in Earth & Environmental Sciences Area (EESA), Lawrence Berkeley National Laboratory (LBNL), Berkeley, CA

Mohamed F. Dahab (water reuse; pollution prevention; sustainable systems for water quality improvement including biological treatment; nutrients removal; biosolids and energy management; use of natural systems for wastewater treatment), College of Engineering, Civil Engineering, University of Nebraska – Lincoln (UNL), Lincoln, NE

Benny D. Freeman (membrane science; oxidatively stable desalination and forward osmosis membrane materials; bioinspired membranes to control fouling of water purification membranes; physical aging of glassy polymeric materials and membranes), Richard B. Curran Centennial Chair in Engineering, Department of Chemical Engineering, University of Texas at Austin (UT Austin), Austin, TX

Pierre Glynn (groundwater geochemistry and integrated modeling; behavioral biogeosciences & policy; green urban infrastructure; multi-resources analyses; ecosystem services; citizen science), Branch Chief for National Research Program in Hydrology; Water Mission Representative to USGS Science and Decisions Center, US Geological Survey (USGS), Reston, VA

James Kilduff (application of sorption and membrane separations to water purification), Department of Civil and Environmental Engineering, School of Engineering, Rensselaer Polytechnic Institute (RPI), Troy, NY

Thomas R. Kulp (microbiological cycling of environmentally relevant trace metal(loids); novel microbial metabolisms involving metalloids that permit bacterial life at extreme conditions; microbiologically mediated reactions), Department of Geological Sciences and Environmental Studies, Binghamton University (BU), Binghamton, NY

Dina L. López (hydrogeology; water quality; environmental geochemistry; geochemistry of hydrothermal systems; fluid flow and heat transfer; acid mine drainage (AMD) and arsenic contamination in water; environmental causes of chronic kidney disease of unknown etiology), Department of Geological Sciences, Ohio University, Athens, OH

Lena Ma (phytoremediation and stabilization of contaminated soil and water), Soil & Water Science Department, University of Florida (UF), Gainesville, FL

Rabi Mohtar (quantify the interlinkages of the water-energy-food nexus that is constrained by climate change and social, political, and technological pressures; design and evaluation of international sustainable water management programs to address water scarcity), Founding Director of Qatar Environment and Energy Research Institute (QEERI); TEES endowed professor, Department of Biological and Agricultural Engineering and Zachary Civil Engineering, Texas A&M University (TAMU), College Station, TX

Lee Newman (phytoremediation; plant nanoparticle interactions; phytoremediation of groundwater contaminated with industrial organic pollutants; hyperspectral imaging of plants to determine contaminant exposure and plant microbe interactions to increase plant productivity and decrease stress responses), Past President, International Phytotechnology Society; College of Environmental Science and Forestry, State University of New York (SUNY), Syracuse, NY

Bethany O'Shea (geogenic contaminants in groundwater; water-rock interactions), Department of Environmental and Ocean Sciences, University of San Diego (USD), Alcala Park, San Diego, CA

Faruque Parvez (human exposure and consequences to geogenic trace contaminants), Department of Environmental Health Sciences, Mailman, School of Public Health, Columbia University (CU), New York, NY

Madeline Schreiber (chemical and contaminant hydrogeology: environmental fate of trace elements; biodegradation processes; epikarst hydrology and geochemistry), Department of Geosciences, Virginia Polytechnic Institute and State University (Virginia Tech), Blacksburg, VA

Arup K. Sengupta (water and wastewater treatment: preparation, characterization and innovative use of novel adsorbents, ion exchangers, reactive polymers and specialty membrane in separation and control; hybrid separation processes), Department of Civil and Environmental Engineering & Department of Chemical Engineering, Lehigh University (LU), Bethlehem, PA

Shikha Sharma (stable isotope geochemistry with focus on issues related to the water-energy-environment nexus and isotope applications in unconventional and sustainable energy resources), Director, WVU Stable Isotope Laboratory, Department of Geology & Geography, West Virginia University (WVU), Morgantown, WV

Subhas K. Sikdar (water quality regulations; clean and sustainable technologies; environmental policy; measurement of sustainability; chemical engineering; water-energy and related nexuses; process development; separation processes), Associate Director for

Science, National Risk Management Research Laboratory, U.S. Environmental Protection Agency, Cincinnati, OH

Vijay P. Singh (surface water and groundwater hydrology; groundwater and watershed modeling; hydraulics; irrigation engineering; environmental quality and water resources; hydrologic impacts of climate change), Distinguished Professor and Caroline & William N. Lehrer Distinguished Chair in Water Engineering, Department of Biological & Agricultural Engineering (College of Agriculture and Life Sciences), Texas A&M University (TAMU), College Station, TX

Shane Snyder (identification, fate, and health relevance of emerging water pollutants and contaminants of emerging concern (ECEs)), Department of Chemical and Environmental Engineering, College of Engineering, University of Arizona (UA); Co-Director of the Arizona Laboratory for Emerging Contaminants, Tucson, AZ

Paul Sullivan (water-energy-food nexus; water and conflict; economics of water resources; water-insurgency-revolution-war nexus; the political economy of water), National Defense University (NDU), Georgetown University; National Council on US-Arab Relations, and Federation of American Scientists, Washington, DC

Paul Sylvester (drinking water; wastewater and nuclear waste treatment; ion exchange; materials chemistry; radiochemistry; product commercialization; research management), Consultant, Waltham, MA

Maya A. Trotz (sustainability; water quality, ecotourism and small scale mining impacts on sustainable livelihoods; climate change; environmental engineering education in formal and informal settings in the US and in developing countries), Civil and Environmental Engineering Department, University of South Florida (USF), Tampa, FL

Richard Tsang (sludge treatment), Senior Vice President, CDM Smith, Raleigh, NC

Nikolay Voutchkov (desalination and water reuse – applied research; advanced technology development and expert review; project scoping, feasibility analysis, planning, cost-estimating, design, contractor procurement and implementation oversight; operation, optimization and troubleshooting for desalination plants), President of Water Globe Consulting, Stamford, CT

Zimeng Wang (water quality; water chemistry; biogeochemistry; water treatment; environmental engineering; soil and groundwater; contaminant transport; geochemical modeling), Department of Civil and Environmental Engineering, Stanford University, Stanford, CA

Y. Jun Xu (ecohydrology; watersheds hydrologic and biogeochemical processes; riverine sediment and nutrient transport; water quality; impacts of land use and global climate change on hydrologic and biogeochemical cycles), School of Renewable Natural Resources, Louisiana State University (LSU), Baton Rouge, LA

Yan Zheng (water, sanitation, and hygiene; hydrochemistry; biogeochemistry of chemical compounds and elements in the environment and their effects on human and ecosystem health; arsenic in groundwater and mitigation), Queens College, City University of New York (CUNY), Flushing, NY & Lamont-Doherty Earth Observatory, Columbia University (CU), Palisades, NY

VENEZUELA

Ernesto Jose Gonzalez Rivas (reservoir limnology; plankton ecology; eutrophication of water bodies), Venezuelan Focal Point for the water program of the InterAmerican Network of Academies of Sciences (IANAS); Institute of Experimental Biology, Faculty of Sciences, Central University of Venezuela (UCV), Caracas

VIETNAM

Vo Thanh Danh (water resources and environmental economics), School of Economics and Business Administration, Cantho University (CTU), Ninh Kieu District, Cantho City

YEMEN

Hussain Al-Towaie (solar power for seawater desalination; thermal desalination), Owner & CEO at Engineering Office "CE&SD" (Clean Environment & Sustainable Development), Aden

ZAMBIA

Imasiku A. Nyambe (geology, sedimentology, hydrogeology, environment and integrated water resources management; remote sensing and GIS; geochemical/environmental mapping; mining), Coordinator, Integrated Water Resources Management Centre, Director of the Directorate of Research and Graduate Studies, University of Zambia (UNZA), Lusaka

ZIMBABWE

Innocent Nhapi (sanitary and environmental engineering; climate change adaptation), Department of Environmental Engineering, Chinhoyi University of Technology (CUT), Chinhoyi

Table of contents

List of contributors

Catherine Aresipathi	Karlsruhe University of Applied Sciences, Karlsruhe, Germany
Pietro Argurio	Department of Environmental and Chemical Engineering, University of Calabria, Arcavacata di Rende (CS), Italy
Jochen Bundschuh	Deputy Vice-Chancellor's Office (Research and Innovation) & Faculty of Health, Engineering and Sciences, University of Southern Queensland, Toowoomba, Queensland, Australia & Royal Institute of Technology, Stockholm, Sweden
Alfredo Cassano	Institute on Membrane Technology, ITM-CNR, Rende (CS), Italy
Franco Cecchi	Department of Biotechnology, University of Verona, Verona & Interuniversity Consortium Chemistry for the Environment, Marghera, Italy
Maria Concetta Carnevale	Institute on Membrane Technology, ITM-CNR, Rende (CS), Italy
João Crespo	LAQV-REQUIMTE, DQ, FCT, Universidade NOVA de Lisboa, Caparica, Portugal
Alessandra Criscuoli	Institute on Membrane Technology, ITM-CNR, Rende (CS), Italy
Shamim-Ahmed Deowan	Karlsruhe University of Applied Sciences, Karlsruhe, Germany
Francesco Fatone	Department of Biotechnology, University of Verona, Verona & Interuniversity Consortium Chemistry for the Environment, Marghera, Italy
Alberto Figoli	Institute on Membrane Technology, ITM-CNR, Rende (CS), Italy
Clàudia Fontàs	Department of Chemistry, University of Girona, Girona, Spain
Noreddine Ghaffour	Water Desalination & Reuse Centre, King Abdullah University of Science and Technology (KAUST), Saudi Arabia
Mattheus Goosen	Office of Research and Graduate Studies, Alfaisal University, Riyadh, Saudi Arabia
Tao He	Laboratory for Membrane Materials and Separation Technology, Shanghai Advanced Research Institute, Chinese Academy of Sciences, Shanghai, China
Jan Hoinkis	Karlsruhe University of Applied Sciences, Karlsruhe, Germany
Evina Katsou	Department of Biotechnology, University of Verona, Verona, Italy & Department of Mechanical, Aerospace and Civil Engineering, Brunel University, London, United Kingdom
Xue-Mei Li	Laboratory for Membrane Materials and Separation Technology, Shanghai Advanced Research Institute, Chinese Academy of Sciences, Shanghai, China
Simos Malamis	Department of Biotechnology, University of Verona, Verona, Italy & Department of Water Resources and Environmental Engineering, National Technical University of Athens, Athens, Greece

Prasanta Kumar Mohapatra	Radiochemistry Division, Bhabha Atomic Research Centre, Trombay, Mumbai, India
Raffaele Molinari	Department of Environmental and Chemical Engineering, University of Calabria, Arcavacata di Rende (CS), Italy
Anh Thi Kim Tran	Department of Chemical Engineering, Process Engineering for Sustainable Systems (ProCESS), University of Leuven, Leuven, Belgium & Faculty of Chemical and Food Technology, HCM University of Technical Education, Vietnam
Spas D. Kolev	School of Chemistry, The University of Melbourne, Melbourne, Victoria, Australia
Hacene Mahmoudi	Faculty of Technology, Hassiba Benbouali University, Chlef, Algeria
Tiziana Marino	Institute on Membrane Technology, ITM-CNR, Rende (CS), Italy
Abdolreza Mirmohseni	Department of Applied Chemistry, Faculty of Chemistry, University of Tabriz, Iran
Priyanka Mondal	Department of Chemical Engineering, Process Engineering for Sustainable Systems (ProCESS), University of Leuven, Leuven, Belgium
Adrian Oehmen	LAVQ-REQUIMTE, DQ, FCT, Universidade NOVA de Lisboa, Caparica, Portugal
Priyanath Pathak	Radiochemistry Division, Bhabha Atomic Research Centre, Trombay, Mumbai, India
Maria Reis	REQUIMTE/CQFB, Chemistry Department, FCT, Universidade Nova de Lisboa, Caparica, Portugal
Stefan-André Schmidt	Institute on Membrane Technology, ITM-CNR, Rende (CS), Italy and Karlsruhe University of Applied Sciences, Karlsruhe, Germany
Mir Saeed Seyed Dorraji	Department of Chemistry, Faculty of Science, University of Zanjan, Iran
Qianhong She	School of Civil and Environmental Engineering, Nanyang Technological University, Singapore & Singapore Membrane Technology Centre, Nanyang Technological University, Singapore
Jianfeng Song	Laboratory for Membrane Materials and Separation Technology, Shanghai Advanced Research Institute, Chinese Academy of Sciences, Shanghai, China
Alexander M. St John	School of Chemistry, The University of Melbourne, Melbourne, Victoria, Australia
Chuyang Y. Tang	School of Civil and Environmental Engineering, Nanyang Technological University, Singapore & Singapore Membrane Technology Centre, Nanyang Technological University, Singapore
Claudia Ursino	Institute on Membrane Technology, ITM-CNR, Rende (CS), Italy
Bart Van der Bruggen	Department of Chemical Engineering, Process Engineering for Sustainable Systems (ProCESS), University of Leuven, Leuven, Belgium
Vahid Vatanpour	Faculty of Chemistry, University of Kharazmi, Tehran, Iran
Svetlozar Velizarov	LAVQ-REQUIMTE, DQ, FCT, Universidade NOVA de Lisboa, Caparica, Portugal

Foreword by Subhas K. Sikdar

Commercial applications of semipermeable membranes started with the invention, in the nineteen sixties, of an asymmetric reverse osmosis membrane for the purpose of producing potable water by desalination. Since then much research gave rise to developing newer and newer membranes. Development of various methods and techniques then led to a multitude of industrial and domestic applications. The objective of membrane methods has mostly been to separate water in some pure form from offending agents, be they particulates, ions, or dissolved or suspended organic chemicals. However, some of the most successful applications have been in separating gases, for instance, nitrogen from air by hollow fibers. The traditional membrane applications rely on the ability of membranes to act as barriers, allowing passage through the membranes based on size of the transferred entities, such as particulates or dissolved ions. In this sense, membranes can be looked upon as an option for separating unwanted materials that span the entire size spectrum from molecular dimension, measured in Angstrom, to particulates, in millimeters. The corresponding membrane methods based on this principle are reverse osmosis, nanofiltration, ultrafiltration, microfiltration and the like.

Barrier technique, however, has not been sufficient to achieve many significant needs. Thus, in parallel, researchers have attempted to exploit membrane affinity as a property to impart a gatekeeping role to membranes. When the membrane affinity is for the desired entity, such as water or nitrogen, the membrane has to offer high flux because the permeating species is dominant in quantity. The separation factor of the permeating species with respect to the undesirable species has to be satisfactory also. Reverse osmosis for desalination, and the famed Monsanto Prism membrane for air separation, are examples of these features. On the other hand, when the permeating species are the minor constituents, such as an organic solvent in aqueous medium worthy of recovery, separation factor will be the dominant criterion for the economy of the process. Examples are pervaporation of organic chemicals from contaminated water, or liquid membranes for heavy metal recovery, the latter exploiting chelating ability of membranes.

A very promising approach has recently emerged by combining adsorption with membranes for organic or metallic species. In this contraption, membranes are endowed with functionalities that bind the target species by physical or chemical bonds preventing them to transgress through the membrane. Unlike a typical barrier process, the success of this operation depends critically on the capacity of the membrane for the target species, and also on the fouling of the active sites by competing species that are invariably present. These membranes need to be regenerated after the sites are saturated. The result of this type of membrane process is the creation of a concentrated stream containing the targeted species which need to be either recovered in case they are valuable or treated when they are undesirable. There is one exception to this regeneration requirement. Sometimes the physical adsorption is taken advantage of to increase the species concentration inside the membrane such that the permeating species experience a larger driving gradient between the membrane and the receiving medium. An example will be pervaporation of volatile organic compounds (VOCs) from a liquid stream to a condensed phase across the membrane. In physical adsorption, the principle of reversible equilibrium adsorption-desorption principle holds, which means the capacity depends on the strength of the incoming stream. The realizable capacity of these adsorptive membranes will be low for low concentration streams and high for high concentration streams. The adsorption isotherms for physical adsorption suggest what can be achieved, not what will be achieved. Too frequent regeneration can be expensive and unaffordable, thus the realizable equilibrium needs to be considerable. Chemisorption provides an escape from this dilemma. Since the bonds that hold the target species to the membrane are chemical, the same reversible equilibrium principle does not apply. High capacity requirement is still a necessity, as it is directly relevant to cost of operation. Microfiltration membranes offer the best potential for this adsorption-membrane combined technique because the affinity ligands can be deposited inside the pores of the membranes, largely increasing the capacity of capture and allowing higher flux of the effluent.

This book is focused on removing trace elements from wastewater or from water that would be used for potable purposes. Arsenic, uranium, and fluorides are especially examined with various membrane methods, including some adsorptive techniques. Arsenic, typically appearing in water as anionic species, became first a local concern in the nineteen nineties when naturally occurring arsenic in groundwater, used for domestic needs, revealed severe health impacts in Bangladesh and parts of Eastern India. Gradually it became a worldwide concern. In the United States, the Environmental Protection Agency ran a demonstration program in which many technologies from all parts of the world were tested for their efficacy in removing arsenic from ground water. Almost without exception the tested technologies, some of which are already deployed, are adsorption-based. They typically exploit chemical reactions to change the valence of the existing arsenic species before adsorptive removal. Several chapters in the book are devoted to sorption of arsenic, including biosorption in membranes. Arsenic is highly toxic, hence a disposal method is always required after its removal. Like arsenic, fluorides also are naturally occurring in groundwater, and when fluorides are present in excessive amounts in potable water, consumption can lead to dental fluorosis and other bone loss diseases. However, fluoride in tiny amounts also fight dental cavities. In the 1984 World Health Organization guideline, fluoride was recommended to be added to drinking water supplies. In many countries, it is also added to tooth pastes. To avoid fluorosis in many countries, the Governments have banned its presence in drinking water. Thus a robust controversy exists on what to do about fluorides. Several authors in the book offer research results on membrane removal of fluorides from water. Uranium, a radioactive metal, also is found in some groundwater which can, if not removed, be a human health concern. Uranium is mostly not absorbed by the human body, but a part stay in the body and can cause kidney problems because of chemical reactions, not because of radioactivity. Uranium is consumed mostly through foods, not water. Being a very heavy metal, it is amenable to be removed by barrier techniques such as reverse osmosis, or by chemical affinity such as ion exchange. Half a dozen chapters in this book deal with uranium removal by several membrane techniques. Like Fluoride, it is not certain what to do about Uranium on the regulatory front. Research results that are presented in the book on removal from water of both elements will prove to be a resource for researchers if their removal from water resources were mandated by the authorities. This book is not meant to be a general purpose treatise on membranes. Its value is in the specialized knowledge of the applicability of membrane methods to remove arsenic, fluorides and uranium from water.

Subhas K. Sikdar
National Risk Management Research Laboratory
U.S. Environmental Protection Agency
Cincinnati, OH
December 2015

Editors' foreword

Uraniun, arsenic and fluoride are found in many regions around the world, where they make freshwater sources unsuitable for drinking or irrigation without prior treatment. The origin of these contaminants can be natural being released from geogenic sources by their mobilization through natural physico-chemical processes affecting, in particular, groundwater resources often on a regional scale and, most importantly, often co-occuring. Since groundwater is increasingly used for freshwater supply, its treatment to an adequate standard is becoming increasingly important. Uraniun, arsenic and fluoride can also be mobilized from geogenic sources through mining into freshwater resources or can be of purely anthropogenic origin resulting from industrial processes. The possibility of finding appropriate technologies to treat freshwater resources contaminated by uraniun, arsenic and fluoride and to reuse treated industrial and other wastewater for reuse for industrial or agricultural or drinking water purposes is one of the major global challenges of the present century.

This book addresses conventional and novel membrane technologies and their application for removing toxic metal(loid)s and halogens from water, with particular attention devoted to the removal of uranium (U), arsenic (As), and fluoride (F^-). All of these compounds exist in the earth's crust at average levels of between two and five thousands micrograms per kg (parts per million) and all are considered highly toxic to humans with exposure primarily through air, food and water. In order to comply with new maximum contaminant levels, numerous studies have been undertaken to improve established treatments or to develop novel treatment technologies for removing toxic metal(loid)s and fluoride from contaminated surface and groundwaters. Among the available technologies applicable for water treatment, membrane technology has been identified as a promising approach for the removal of such toxic contaminants from water. Therefore, this book is devoted to describing both pressure driven traditional methods (Nanofiltration, Reverse Osmosis, Ultrafiltration, etc.) and more advanced and novel membrane processes (such as Forward Osmosis, Membrane Distillation and Membrane Bioreactors) employed in the removal of uranium, arsenic and fluoride.

One key goal of this book is to provide information both on membrane technologies and on the results obtained in their application. All the authors involved in the writing of the chapters are experts in the specific membrane technology described and this makes the book really unique. The information provided should facilitate the choice of process suitable for a specific application and also show the potential of these innovative membrane processes. The different chapters each cover a specific membrane process, allowing the possibility of describing the technology in detail and evaluating its potentiality.

This book is divided in three parts: Part I contains the introductory Chapter 1 which provides an overview of the chemistry and human concerns of the toxic contaminants (U, As, and F^-) to be removed from water by membrane technologies. Part II contains five chapters dealing with the use of different conventional membranes while Part III gives detailed insights into new trends in materials and membrane process development to be applied for the target contaminants from water. Within Part II, Chapter 2 provides details on the basic principles of microfiltration (MF) and ultrafiltration (UF) processes and selected applications in the treatment of drinking water specifically with respect to arsenic removal. It also illustrates and highlights the significant advantages which can be achieved through the integration of these membrane technologies with adsorption and coagulation/flocculation technologies. These principles can be generally extended to the removal of fluoride and uranium from drinking water. Chapter 3 reports the successful application of nanofiltration in the removal of fluoride and uranium from groundwater in order to meet WHO regulations. Chapter 4 focuses on the reverse osmosis (RO) process and the removal of As, F^- and U by reverse osmosis (RO) using different types of membrane modules. Chapter 5 presents the most relevant electromembrane processes for treating water containing traces of toxic contaminants and addressing the removal of As, F^- and U. Two case-studies are reported and discussed in order to illustrate the possible successful application of Donnan dialysis, applied either as a single treatment or part of an integrated approach for the removal of arsenate and ionic mercury from contaminated water sources. Chapter 6 reports the removal of the three target contaminants from water and wastewater using adsorbent materials (including mixed matrix membrane) and integrated membrane systems. In Chapter 7, the application and comparison of liquid membrane techniques for the removal of U from lean solutions is extensively evaluated and discussed. Chapter 8 reports the main theoretical aspects of transport in supported liquid membranes (SLMs) and their potential applications in the removal and recovery of toxic ions from water with a specific study on arsenic removal. Chapter 9 describes the Polymer Inclusion Membranes (PIMs) process and explores progress made on the development of PIMs for the separation of U and As from aqueous solutions. The separation of U and As illustrates the applicability of PIM-based technology, which is considered to be a relatively novel type of self-supporting liquid membrane with low energy requirements, to industrial separation and environmental remediation. In Chapter 10, the removal of arsenic by means of nanofiltration (NF) membranes is discussed. The aim of this chapter is to give a comprehensive overview of the use of NF membrane-based processes for As removal, providing not only a critical analysis of the current treatment status of using membrane based processes, but also pointing out new development directions for arsenic removal in a more energy efficient manner and reporting a specific case study on the preparation of high performance NF membranes. Chapter 11 reports the positive effect of the coupling of biochemical processes with membranes in membrane bioreactors (MBR). The MBR process is described in detail and its potential for use in water treatment and removal of toxic compounds is also reported. The basic principles of membrane distillation and the application of this membrane process to treat polluted water, in terms of permeate flux and contaminant rejection, are reviewed and discussed in depth in Chapter 12. Finally, in Chapter 13, the potential of Forward Osmosis (FO), considered an emerging membrane technology, for the removal of contaminants (both organics and inorganics), particularly As, from impaired water is discussed.

The book is a "first of its kind"; as there are no other contemporary publications on this topic available and we believe that this book will provide the readers with a thorough understanding of the different available membrane technologies for the removal of traces of toxic compounds such as uranium, arsenic, and fluoride from water.

We hope that this book will help all readers, professionals, academics and non-specialists, as well as key institutions that are working on membrane technology and water treatment projects. It will be useful for leading decision and policy makers, water sector representatives and administrators, policy makers from government, business leaders, companies involved in water treatment, and engineers/scientists from both industrialized and developing countries. It is expected that this book will become a standard, used by educational institutions and Research and Development establishments involved in the respective issues.

Alberto Figoli
Jan Hoinkis
Jochen Bundschuh
(editors)
November 2015

About the editors

Alberto Figoli (1970, Italy) got his PhD degree at Membrane Technology Group, Twente University (NL) in 2001. He is Senior Researcher at the Institute on Membrane Technology (ITM-CNR) in Rende (CS), Italy, since 2001.

Alberto Figoli is expert in the field of membrane technology, particularly in *membrane preparation and characterisation* and *application in water treatment.*

In 1996, he obtained his Master Degree in Food Science and Technology, at the Agriculture University of Milan. Then, he worked for about 1 year at Quest International Nederland B.V. (ICI), Naarden (The Netherlands) at the Process Research Group, on a pilot plant for aromatic compounds extraction using the pervaporation membrane technology. He has been granted twice by National Research Council of Italy (CNR) to the Short Term Mobility Programme, as visiting researcher of the Environmental Protection Agency of United States (USEPA) at the Sustainable Technology Division in Cincinnati (USA). The research was devoted to pervaporation studies in the environmental field.

In recent years, he has been involved in several European and National projects (as scientist responsible or principal investigator for ITM-CNR) in the field of Membrane Technology. In particular, he has been the scientific coordinator of several National projects with Industries and a Marie Curie EU Project. He is the author of more than 80 scientific peer-review papers and chapters published in international journals and books. He is the editor of one book and author of two patents on membrane technology.

In 2015, he was elected as Council Member of the European Membrane Society, EMS, for the period 2015–2019. He is responsible for the Awards and Summer Schools.

Jan Hoinkis (1957, Germany) conducted a doctorate in thermodynamics at the University of Karlsruhe, Germany (now Karlsruhe Institute of Technology). After completion of his thesis he moved to the Swiss company Ciba-Geigy, where he was working as head of an R&D group on process development in the field of fine chemicals production with focus on environmentally friendly technologies. He has been working since 1996 as a professor at the Karlsruhe University of Applied Sciences giving lectures in chemistry, thermodynamics as well as environmental process engineering. His R&D work is focused on water treatment and water reuse with special attention on sensor-based membrane technologies. He was involved in several national and international R&D projects. In 2008 he was appointed Scientific Director of the Institute of Applied Research, which is the central research facility at the Karlsruhe University of Applied Sciences.

Jochen Bundschuh (1960, Germany), finished his PhD on numerical modeling of heat transport in aquifers in Tübingen in 1990. He is working in geothermics, subsurface and surface hydrology and integrated water resources management, and connected disciplines. From 1993 to1999, he served as an expert for the German Agency of Technical Cooperation (GTZ – now GIZ) and as a long-term professor for the DAAD (German Academic Exchange Service) in Argentina. From 2001 to 2008 he worked within the framework of the German governmental cooperation (Integrated Expert Program of CIM; GTZ/BA) as adviser in mission to Costa Rica at the Instituto Costarricense de Electricidad (ICE). Here, he assisted the country in evaluation and development of its huge low-enthalpy geothermal resources for power generation. Since 2005, he has been an affiliate professor of the Royal Institute of Technology, Stockholm, Sweden. In 2006, he was elected Vice-President of the International Society of Groundwater for Sustainable Development ISGSD. From 2009–2011 he was visiting professor at the Department of Earth Sciences at the National Cheng Kung University, Tainan, Taiwan.

Since 2012, Dr. Bundschuh has been professor in hydrogeology at the University of Southern Queensland, Toowoomba, Australia where he is working in the wide field of water resources and low/middle enthalpy geothermal resources, water and wastewater treatment and sustainable and renewable energy resources. In November 2012, Prof. Bundschuh was appointed as president of the newly established Australian Chapter of the International Medical Geology Association (IMGA).

Dr. Bundschuh is author of the books "Low-Enthalpy Geothermal Resources for Power Generation" (2008) (Taylor & Francis/CRC Press) and "Introduction to the Numerical Modeling of Groundwater and Geothermal Systems: Fundamentals of Mass, Energy and Solute Transport in Poroelastic Rocks". He is editor of 16 books and editor of the book series "Multiphysics Modeling", "Arsenic in the Environment", "Sustainable Energy Developments" and the recently established series "Sustainable Water Developments" (all CRC Press/Taylor & Francis). Since 2015, he has been editor in chief of the Elsevier journal "Groundwater for Sustainable Development".

Acknowledgements

The editors thank the authors of the chapters for their participation, discussion and contribution. The editors would like to remember one of the authors, Dr. Priyanath Pathak from India, who left us prematurely, for his devotion to research. The editors and authors thank also the technical people of Taylor & Francis Group, for their cooperation and the excellent typesetting of the manuscript.

Part I
Generality on arsenic, fluoride and uranium

CHAPTER 1

Fluoride, uranium and arsenic: occurrence, mobility, chemistry, human health impacts and concerns

Alberto Figoli, Jochen Bundschuh & Jan Hoinkis

1.1 INTRODUCTION

Mainly due to global population growth the demand for potable water is continuously rising. According to the United Nations, world population is projected to reach 9.6 billion by 2050 with most growth in developing regions, especially in Africa (UN, 2013). During the same period, the population of developed regions will remain largely unchanged at around 1.3 billion people (UN, 2013). This forecast highlights well the need for safe drinking water especially in less developed regions, where the population is expected to grow significantly. About 97% of the freshwater reserve is stored in aquifers, which makes the groundwater the largest global freshwater resource. That resource caters to the need of a population of over 1.5 billion (Jacks and Battacharya, 2009). When comparing this freshwater to surface water quality many advantages turn out. Groundwater is generally free from pathogenic bacteria and viruses and has far lower concentrations of organic matter. However, a variety of organic and inorganic contaminants have been identified in groundwater that are potentially toxic to humans or animals (Hoinkis *et al.*, 2011). The origin of these contaminants is on the one hand naturally occurring through mobilization from the rocks and minerals through physical, chemical, and microbiological processes into the groundwater while on the other hand uniquely human sources like pesticides, fertilizers or industrial and mine waste discharge are other sources.

Rapid and intensive industrialization has generated large volumes of aqueous wastes containing dangerous materials, such as heavy metals and metalloids. Water contamination by heavy metals, metalloids and other minor and trace elements such as fluoride constitute a big global health hazard (An *et al.*, 2001; Mulligan *et al.*, 2001) as they can be toxic and carcinogenic even at very low concentrations, and, hence, usually pose a serious threat to the environmental and public health (Liu *et al.*, 2008; Vilar *et al.*, 2007). During traditional wastewater treatment, most heavy metals (e.g., lead, chromium and cadmium) and metalloids (e.g., arsenic (As)) pass unhindered through the treatment process, which is mainly due to their occurrence in trace amounts. In fact, little to no attention has historically been given to metals and metalloids in wastewater treatment plants.

Natural sources (volcanic emission, weathering of rocks and microbiological activity), release of geogenic contaminants through mining and anthropogenic sources (e.g., burning of fossil fuels, use of arsenical pesticides and herbicides, etc.) are responsible for high As concentrations in water in many parts of the world (ATSDR, 2007). In the affected areas, As concentrations in groundwater are generally found in the range of 100–2000 $\mu g\,L^{-1}$ On the other hand, the potential sources for high concentration of F^- in water are dissolution of F^- bearing rocks under favorable natural conditions and/or discharge of F^- contaminated wastes from the industry (Mohapatra *et al.*, 2009). Moreover, like As and F, U is also distributed in the environment due to natural (weathering of rock) and anthropogenic (mining, nuclear power production and phosphate fertilization) sources and leaves a very high impact on the environment, which is a latent risk factor for both human and animals (Langmuir, 1997; Oliver *et al.*, 2008).

As a result of the high concentration of these chemical species in groundwater, an adequate treatment is required for removal of these contaminants before supplying it for human consumption.

Various chemical treatment technologies have been applied to remove these ions from drinking water sources, including ion exchange, metal oxide based adsorption and coagulation. However, these methods alone are insufficient to remove the contaminants below the Maximum Contaminant Limit (MCL) and therefore are better to be used as a pretreatment step (Favre-Réguillon *et al.*, 2005; Mondal *et al.*, 2013).

In fact, the presence of such inorganic arsenic (As(V/III)), fluoride (F^-) and uranium (U(VI)) species (mostly ions) in groundwater (and to less extent in surface water) is a critical global issue, and has created severe health impacts for decades. Bioaccumulation and adverse effects on human health by intake of these ions via drinking water have been well documented (e.g., Fawell *et al.*, 2006; Orloff *et al.*, 2004; Smedley and Kinniburgh, 2002).

The aim of this book is to describe, analyze and bring to the attention the existence of different types of membrane processes, which could be successfully applied for the removal of toxic metals from water. In particular, the removal of As, U and F^- fluoride will be taken into consideration as specific cases.

This introductory chapter for this volume provides basic information on the occurrence and chemical species of As, U and F^- in freshwater resources (predominantly groundwater), their release from rocks and sediments and mobility as well as the principal health impacts, which occur due to human uptake through drinking water or through the human food chain. The chapter provides only simplified insight into these topics as far as this knowledge is needed for selecting the most appropriate technology and design for removal of these trace contaminants from drinking and irrigation water and provide the reader with knowledge of the global importance of contamination of freshwater resources with these geogenic and, of minor importance anthropogenic, trace elements and related potential health impacts, which clearly demonstrate the importance of their removal through adequate treatment.

1.2 FLUORIDE

Fluorine is the lightest, reactive and most electronegative element in the halogen group of the periodic system and has a strong tendency to acquire a negative charge. Thus, it remains as a negative ion (F^-) in solution (Fawell *et al.*, 2006) and forms negative and positive complexes (e.g., dissolved $[MgF]^+$ complexes). Soluble fluoride complexes with Al^{3+}, Fe^{3+} and Si^{4+} have high equilibrium constants ranging from 10^6 to 10^5, but the amount of Al^{3+}, Fe^{3+} and Si^{4+} ions is below $1\,mg\,L^{-1}$ in most natural waters (pH 5 to 8, Eh $= -200$ to $+200\,mV$) (Baas Becking *et al.*, 1960). Graham *et al.* (1975) and Roberson and Barnes (1978) state that fluoride complexes with Al^{3+}, Fe^{3+} and Si^{4+} must therefore only be considered at rather low pH-values. In areas where high fluoride concentrations are correlated with arsenic like in Arizona (Robertson, 1984) and especially in the Argentine Pampa and Chaco plains, additionally the fluoride complexes of As must be considered (e.g., $HAsO_3F^- + H_2O = F^- + H^+ + H_2AsO_4^-$, $pK = -46.112$ and $AsO_3F^{2-} + H_2O = F^- + H_2AsO_4^-$, $pK = -40.245$) (Bundschuh *et al.*, 2000; 2004). However, ionized and non-ionized organic and inorganic F occur in the environment.

1.2.1 *Sources, release and mobility*

In many regions, fluorine is a widely distributed constituent found in sedimentary porous aquifers, in porous aquifers formed by the overburdens of hard bedrock aquifers and in hard rock aquifers in concentrations beyond the WHO guideline value of $1.5\,mg\,L^{-1}$ (Fawell *et al.*, 2006).

The presence of F^- in the environment occurs not only naturally through its presence in the earth's crust but also due to industrial activities, such as electroplating, semiconductor manufacturing, glass making, steel production and fertilizer industries (Sujana *et al.*, 1998; Toyoda and

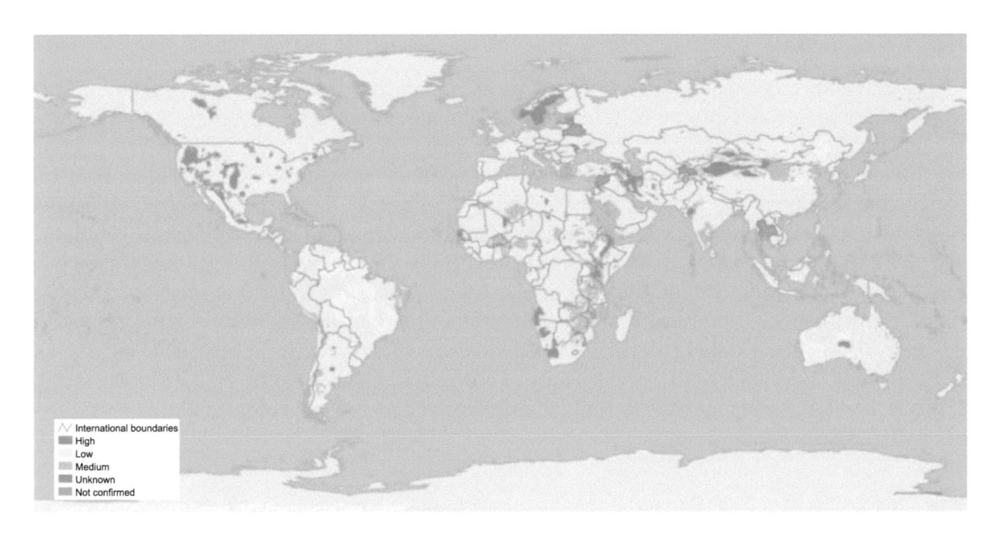

Figure 1.1. Fluoride in groundwater – Probability of occurrence (IGRAC, 2014 Environmental Data Explorer, compiled from IGRAC: Internet Site: http://geodata.grid.unep.ch/options.php?selectedID=2241&selectedDatasettype=16 (accessed on 05 February 2014).

Taira, 2000). The release of wastewater from these industries leads to the F^- pollution of surface and groundwater. The US Environmental Protection Agency (USEPA) established the effluent discharge standard of 4 mg L^{-1} for F^- from a wastewater treatment plant (Khatibikamala *et al.*, 2010; Shen *et al.*, 2003). The breakdown of rocks and soils or weathering and deposition of atmospheric volcanic particles are the biggest source of F^- found in the groundwater. Because they often contain abundant F^--bearing minerals crystalline rocks, especially granites, are particularly susceptible to F^- build-up. Concentrations of F^- in groundwater can range from below 1 mg L^{-1} to more than 50 mg L^{-1}.

Figure 1.1 provides an overview over the probability of occurrence of the F^- distribution worldwide. The known hotspots of F^- with high concentrations in groundwater are found to be in Scandinavia, China, Western India, East Africa North America and South America. Despite this forecast a variety of different publications showed several other spots of high F^- concentrations in groundwater. In Pakistan, Thailand, China, Sri Lanka, eastern and southern Africa high groundwater F^- concentrations associated with igneous and metamorphic rocks such as gneisses and granites have been reported (Fawell *et al.*, 2006). About half of the states and territories in India reported to have naturally high concentrations of F^- in water (UNICEF, 1999). In Sri Lanka, concentrations up to 10 mg L^{-1} have been reported and in China fluorosis has been reported to be widely spread (Fawell *et al.*, 2006). Since millions of people worldwide are exposed to high F^- concentrations this poses a serious global health threat to the consumers.

The northern Tanzania region is known for being among the most F^- affected areas worldwide. Already in the early 1980s, Nanyaro *et al.* (1984) reported high F^- contents in some rivers, springs, alkaline ponds and lakes in northern areas of Tanzania. The F^- contents found have been 12–26 mg L^{-1} for rivers, 15–63 mg L^{-1} for springs and even mg 60–690 L^{-1} for alkaline ponds and lakes. Also Bugaisa (1971) identified particular F^- problems in groundwater of Tanzania. The concentrations found vary between 4 and 330 mg L^{-1}. Such concentrations are extremely high compared to other F^- contaminated groundwater sources. In the East-African Rift zone lavas (intrusions and ashes) and other volcanic rocks with fluorine-rich minerals are found in much higher concentrations than in similar rock types elsewhere in the world (Kilham and Hecky, 1973). Hot springs are also an important source for high F^- concentrations in the groundwater. In addition, in extreme cases of evaporation of lakes coexisting with infiltration of lake water to the shallow aquifers, F^- contamination of the aquifer might occur.

1.2.2 *Human health effects*

Fluoride toxicity can happen by a number of ways. Bhatnagar *et al.* (2011) noted that the impact of F^- in drinking water can be beneficial or detrimental to human well-being. Small amounts in consumed water, for example, are usually considered to have a beneficial effect by reducing the rate of occurrence of dental cavities, predominantly amongst children (Mahramanlioglu *et al.*, 2002). In contrast, consumption of large amounts of F^- has been shown to lead to diseases such as osteoporosis, arthritis, cancer, infertility, brain damage, Alzheimer syndrome, and thyroid disorder (Chinoy, 1991; Harrison, 2005). In addition, F^- has been shown to poison kidney function at high doses over short-term exposures in both animals and humans (http://www.fluoridealert.org/studies/kidney01/). Fluoride exposure has also been linked to bladder cancer particularly among workers exposed to excess F^- in the workplace (Bhatnagar *et al.*, 2011; Chinoy, 1991). Thyroid activity is also known to be influenced by F^- (Harrison, 2005). There is therefore an urgent need to find out effective and robust technologies for the removal of excess F^- from drinking water.

1.3 URANIUM

Due to growing global energy demands, nuclear power appears to be a long term prospect as an alternative to fossil fuel based power sources. At this moment, most of the nuclear plants operate on enriched uranium (U) based fuels, thus making U one of the most precious elements.

1.3.1 *Sources, release and mobility*

The sources of U are commonly known minerals such as uraninite (UO_2) and pitch blend (U_3O_8); though other minerals such as carnotite, autunite, uranophane, torbernite, and coffinite also contain U. Secondary sources of U include phosphatic rocks and minerals such as lignite and monazite. Estimated U deposit in phosphatic rocks (world average U content in phosphate rock is estimated at 50–200 mg kg^{-1}) is about 9 million tons (http://www.wise-uranium.org/uod.html). The U recovery from the minerals involves acid (using dilute sulfuric acid) or alkaline (using sodium carbonate) leaching followed by solvent extraction methods using D2EHPA (di-2-ethylhexylphosphoric acid) and tri-alkyl amines which are well known as the DAPEX (dialkylphosphoric acid extraction) and AMEX (amine extraction) processes, respectively. Uranium produced through these processes is further purified to obtain nuclear-grade U by a solvent extraction method from a nitric acid medium using TBP (tri-*n*-butyl phosphate) in kerosene as the extractant.

The metal can also occur in the environment as a result of nuclear industry activities, mill tailings and fuels combustion. Its chemical toxicity is even greater than its radioactivity (Grenthe *et al.*, 1992; Sheppard *et al.*, 2005). The maximum admissible concentration (MAC) of U for drinking water according to the USEPA is 30 $\mu g L^{-1}$. The the World Health Organization (WHO) which introduced in 1998 the health-based drinking-water guideline for U of 2 $\mu g L^{-1}$ but increased to a 30 $\mu g L^{-1}$ in 2011 (WHO, 2004; Reimann and Banks, 2004). Figure 1.2 shows the U reserves in the world (OECD, 2010).

The map of Figure 1.2 highlights where the known natural sources of the metal are significant or low. Uranium is found in ground- and surface waters due to its natural occurrence in geological formations. Countries like Australia and Kazakhstan appear to be U-rich (with reserves of about 1,673,000 and 480,300 metric tons, respectively) and relatively high concentrations of U in groundwater are expected in these cases. Elevated levels of U in water can be also found outside the areas shown in the world map, however. The presence of U in groundwater has been reported, for instance, also in Europe (Scandinavia, Sweden, Finland, Spain, Portugal, France, the Czech Republic and Ukraine) (Raff and Wilken, 1999). The formation of U deposits is essentially a normal geological process leading to its presence in granite and sedimentary rocks. Naturally occurring U is located only in minor amounts on dry land, though. In fact, the estimated quantity

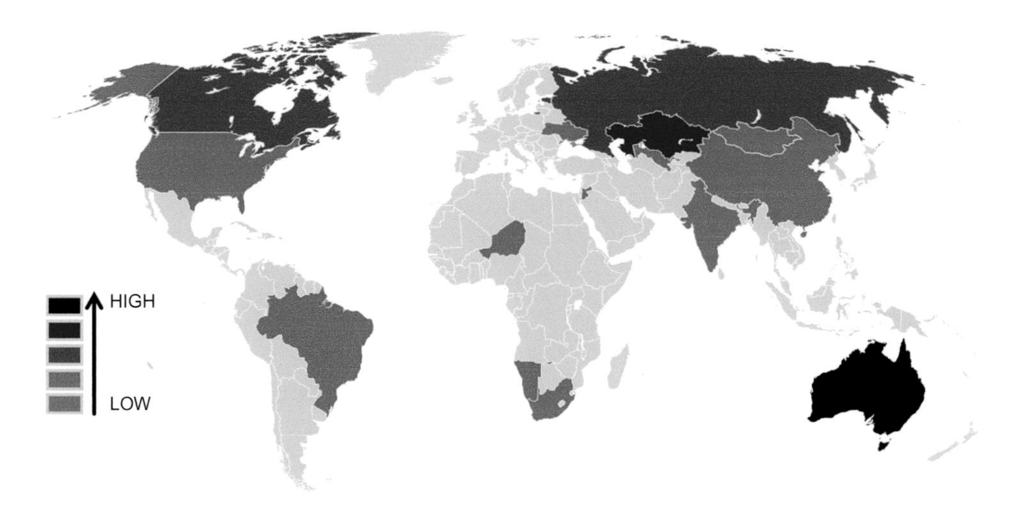

Figure 1.2. Uranium reserves in the world (OECD, 2010).

of dissolved U (3–4 $\mu g\,L^{-1}$) in seawater is ~4.5 billion metric tons, which is approximately 1000 times greater than that of U resources available in different known rock formations suitable for U exploitation (OECD, 2000). In view of this, seawater is being considered as a potential source of U. However, the process should be effective and economical for the selective concentration of uranyl ions. Japan has carried out extensive efforts for the recovery of U from seawater and several materials/processes have been developed and proposed toward the achievement of this goal (Kanno, 1984; Kawai *et al.*, 2008; Kobuke *et al.*, 1988; Schwochau, 1984).

1.3.2 *Human health effects*

Uranium occurs naturally in variable concentrations in all soils, minerals, rocks and waters. It can also be derived from several anthropogenic sources. Uranium is weakly radioactive and human exposure to the element has long been considered to pose a radiological as well as toxic hazard (WHO, 2004; 2008; Smedley *et al.*, 2006). This dissolved toxic radioactive metal may poison drinking water sources and the food chain via contaminated surfaces and groundwater. In recent years, there has been increasing concern that the chemical effects of uranium may also pose a potential hazard to exposed populations. However, there are few if any epidemiological studies that have been able to demonstrate any resultant harm, even in occupational contexts (The Royal Society, 2001).

The primary non-carcinogenic toxic effect of uranium is on the kidneys. Published studies in rats, rabbits, and humans show effects of chronic uranium exposure at low levels in drinking water. Effects seen in rats, at the lowest average dose of $0.06\,mg\,U\,kg^{-1}\,day^{-1}$, including histopathological lesions of the kidney tubules, glomeruli and interstitium are considered clearly adverse effects albeit not severe (OEHHA, 2001).

However, little information is available on the chronic health effects of exposure to environmental uranium in humans which makes it difficult to establish adequate guideline and regulatory limits for uranium in drinking water. Human risk from drinking water is significantly higher due to chemical exposure compared to radiation; the last would only be significant if uranium concentrations exceed $100\,\mu g\,L^{-1}$ (WHO, 2004).

In drinking water, histopathological effects were also seen at the same exposure level in the liver including nuclear anisokaryosis and vesiculation. Effects on biochemical indicators of kidney function were seen in the urine of humans exposed to low levels of uranium in drinking water for periods up to 33 years. Uranium is an emitter of ionizing radiation, and ionizing radiation

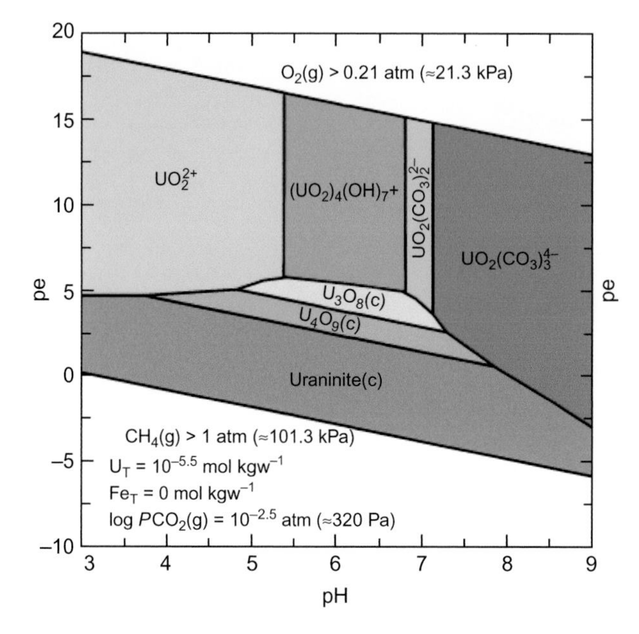

Figure 1.3. pe/pH predominance diagrams at 10°C for uranium (adpated from The Royal Society, 2001). (kgw: kg water).

is carcinogenic, mutagenic and teratogenic. A level of 0.5 μg L^{-1} (0.43 pCi L^{-1}) is considered protective for both carcinogenicity and kidney toxicity (OEHHA, 2001).

1.3.3 *Uranium species*

The mobility of U in water is controlled by a number of factors, among the most important being pH, redox status and concentrations of coexisting dissolved ions. Uranium occurs in the environment in several oxidation states (+2, +3, +4, +5 and +6), but U(IV) (uranous) and U(VI) (uranyl) are most common in the natural environment. In oxic groundwater with low pH, UO_2^{2+} is predominant (Langmuir, 1997), whereas U(IV) is mainly present in reducing groundwater environment and in solid phase (Fig. 1.3) (Rossiter *et al.*, 2010). Complexes of UO_2^{2+} with carbonate, oxalate, and hydroxide are formed in the aqueous medium (Favre-Reguillon *et al.*, 2003). In addition, the UO_2^{2+} cation gets easily sorbed on different surfaces with the aid of dissolved organic matter or by complexation/precipitation with several anions such as hydroxide, silicates, phosphates etc. Uranyl ions are readily soluble and are easily transportable via forming complexes with carbonate (Zhou and Gu, 2005), phosphate (Cheng *et al.*, 2004), hydroxides and organic matters and these complexes are redundant in groundwater with basic pH (Rossiter *et al.*, 2010). The pE (potential)/pH diagram illustrates the different U species in groundwater.

1.4 ARSENIC

Arsenic (As) is a naturally occurring element present in food, water, and air. This element has been known for centuries to be an effective poison. Arsenic, a semi-metal element in the periodic table, is odorless and tasteless. It enters drinking water supplies from natural deposits in the earth or from agricultural and industrial practices. Since groundwater contamination by geogenic arsenic is known from over 70 countries (Fig. 1.4) and because As-associated human health problems have now been recognized in many parts of the world, mainly in developing countries, it is a problem and challenge of global concern.

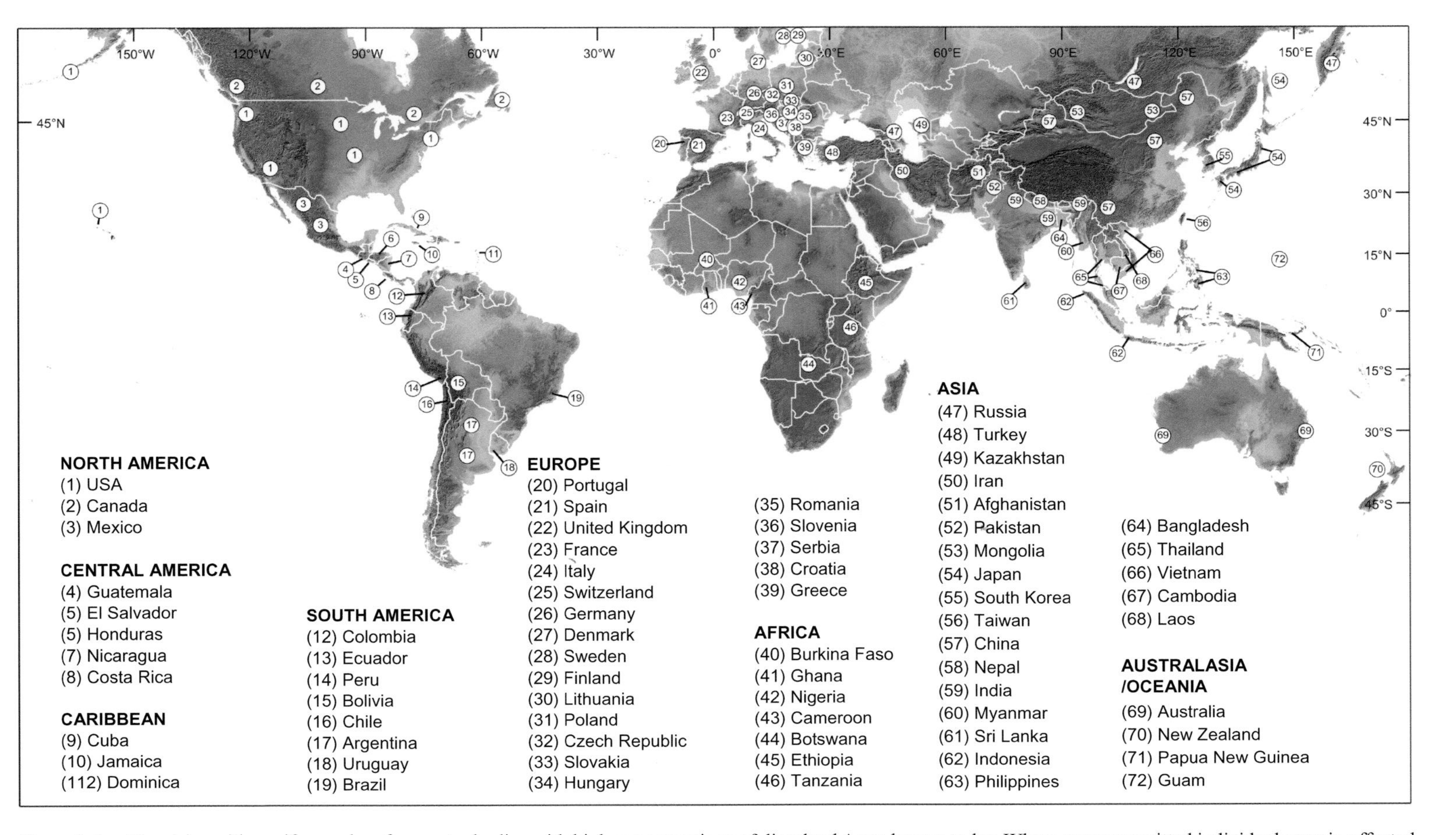

Figure 1.4. Countries with aquifers and surface water bodies with high concentrations of dissolved As as known today. Where space permitted individual arsenic-affected areas are shown within the individual countries. Adapted from Bundschuh and Litter (2010) based on compilations and data from: Bundschuh *et al.*, 2008a; Chandrasekharam and Bundschuh, 2008; Ravenscroft, 2007; Smedley, 2006; Welch *et al.*, 2000; 2009a, and from unpublished information of the authors.

In order to understand the global importance of the problem of high arsenic (As) concentrations in water used for drinking and irrigation water supplies, and the required mitigation needs, in the following sections a short global overview of the occurrence of geogenic As in water resources and a characterization of the conditions under which As is released from the solid aquifer material of from sediments into groundwater and surface water bodies, which is helpful for the formulation of sustainable mitigation solutions, such as As removal from drinking water or zero-treatment options (Section 1.4.1). To show the importance of As removal from water, in the following section (1.4.2) a short overview on human health effects due to exposure to As from water sources is provided. A last section provides information of arsenic species as the species in specific types of groundwater is essential for the selection and design of the As removal technology.

1.4.1 *Groundwater arsenic: sources, release and mobility*

Toxic As concentrations in groundwater and to lesser extent in surface water mostly result from physical or chemical abiotic and biotic weathering of primary or secondary As-containing minerals. Therefore, aquifers with elevated levels of groundwater As derived from natural sources can be classified according to the origin of the As and the mobilization mechanism into the water:

- In reduced alluvial aquifers, the As is predominantly released from iron oxyhydroxides: principle examples are the flood and delta plains of the Himalayan rivers, i.e., the Ganges-Brahmaputra-Meghna plain/deltas in India and Bangladesh (Bhattacharya *et al.*, 2002a; 2002b; 2007a; 2007b; Bhowmick *et al.*, 2013; Prakash Maity *et al.*, 2011a; Reza *et al.*, 2011a), the Indus plain in Pakistan (Nickson *et al.*, 2005), the Irrawady delta in Myanmar (WRUD 2001), the Red River delta in Vietnam (Berg *et al.*, 2001), and the Mekong river delta in Cambodia and Laos (Feldman and Rosenboom, 2001).
- In oxidized, mostly sedimentary, aquifers (normal pH range, $5 < pH < 10$) where desorption of As from Fe, Al, and Mn oxyhydroxides occurs at high pH (8–9.5). This is the principal mechanism responsible for the high As concentration of geogenic As in the groundwater of extended areas of the Argentine Chaco-Pampean plain and its continuation into the neighboring countries Uruguay, Paraguay and Bolivia (Bhattacharya *et al.*, 2006; Bundschuh *et al.*, 2004; 2008a; 2009a; 2009b; 2009c; 2012a; Chatterjee *et al.*, 2010; Nicolli *et al.*, 2010; 2012; Raychowdhury *et al.*, 2013; Smedley *et al.*, 2005; 2009), the Appalachian Highlands, NE Ohio (Matisoff *et al.*, 1982); the Interior Plains, S. Dakota, (Carter *et al.*, 1998); the Carson Desert, Nevada, (Welch and Lico, 1998); the Pacific Mountain System, NW Washington (Davies *et al.*, 1991; Ficklin *et al.*, 1989) and Arizona (Goldblatt *et al.*, 1963; Nadakavukaren *et al.*, 1984). In the groundwater of dry regions, As is often positively correlated to fluoride (Alarcon-Herrera *et al.*, 2013).
- In oxidized environments, at very low pH ($pH < 4$), the geogenic As is released by sulfide oxidation. This process is found in many areas with deposits of sulfide mineral ores. Examples include the mineral deposits in the Andes and Andean highlands (Ormachea Muñoz *et al.*, 2013; 2015; Ramos Ramos *et al.*, 2012; 2014), the Middle and North American cordillera, the Transmexican volcanic belt, the Appalachian belts from Massachusetts to Maine (Ayotte *et al.*, 1998; Boudette *et al.*, 1985; Peters, 2008; Peters *et al.*, 1999; Zuena and Keane, 1985), the Interior Plains of E. Michigan, (Westjohn *et al.*, 1998), the Variscian mountains in Europe, the Central Balkan peninsula in Siberia (Dangic and Dangic, 2007), Albania (Lazo *et al.*, 2007), Ghana (Smedley, 1996), Nigeria (Gbadebo, 2005), and many mining sites all around the world. Gold mining forms another geogenic source for contamination of water resources by arsenic as observed e.g. in Brazil (Ono *et al.*, 2012) and Cuba (Toujaguez *et al.*, 2013)

Other mechanisms, which can play a role in controlling As mobilization are the site-specific geomorphological, geological and hydrogeological conditions as well as climate, land use patterns, and groundwater exploitation (Bundschuh *et al.*, 2010; Hasan *et al.*, 2007; Mukherjee *et al.*, 2007) and the inflow of As-rich geothermal water into freshwater sources (Birkle *et al.*, 2010; Bundschuh and Prakash Maity, 2015; Bundschuh *et al.*, 2013; López *et al.*, 2012) or from release of mud volcano fluids mixing with surface water (Jean *et al.*, 2010; Liu *et al.*, 2011; 2013; Prakash Maity *et al.*, 2011b), and the presence and contents and species of organic matter (Jean *et al.*, 2010; Liu *et al.*, 2013; Reza *et al.*, 2011b). Arid and semiarid climate can be a principal or an additional control and, due to evaporative concentration increases, contribute to the genesis of As-rich groundwater and surface water. Examples where climate contributes to increased As concentrations in groundwater and surface waters include the Atacama desert in N Chile (Borgoño and Greiber, 1971; Bundschuh, 2008a; 2008b; 2008c; 2009a; 2009b), aquifers in the Chaco-Pampean plain in Argentina (Bhattacharya *et al.*, 2006; Bundschuh *et al.*, 2000; 2004; 2008a; 2008b; 2008c; 2009b; 2009c; 2012a; Chatterjee *et al.*, 2010; Nicolli *et al.*, 2010; 2012; Raychowdhury *et al.*, 2013), shallow aquifers of the Carson desert in Nevada (Fontaine, 1994; Welch and Lico, 1998) and shallow aquifers in the southern San Joaquin valley in California (Fujii and Swain, 1995; Swartz, 1995; Swartz *et al.*, 1996).

1.4.2 *Human health effects related to arsenic exposure*

Inorganic As is a Class 1, non-threshold, carcinogen, and chronic exposure also causes a range of ailments such as skin lesions (hyperkeratosis, melanosis), nervous system impairment, irritation of respiratory organs and the gastrointestinal tract, anemia, liver disorders, vascular illnesses and even diabetes mellitus, skin, lung and bladder cancer (Albores *et al.*, 2001; Bates *et al.*, 1992; Cullen *et al.*, 1989; Del Razo *et al.*, 2000; 2005; Endo *et al.*, 2003; Kirk and Sarfaraz, 2003; McClintock *et al.*, 2012; Oremland *et al.*, 2003; Rossman, 2003). Chronic As exposure can also affect the intellectual development of children (Borja *et al.*, 2001; Wasserman *et al.*, 2004).

Ground and surface water sources with As concentrations at toxic level, used for drinking and irrigation, poses a direct threat to human health and environmental sustainability. When considering human exposure to inorganic As, besides direct ingestion from drinking water, other exposure pathways include water used in food preparation (such as rice cooking), direct contamination of food sources through the use of groundwater in agricultural irrigation or aquaculture, and the indirect contamination of food sources, such as feeding straw elevated in As to livestock (e.g., Bundschuh *et al.*, 2012b and references therein).

At present, it is not possible to precisely assess the number of persons potentially exposed worldwide to elevated inorganic As from groundwater and surface water exposure routes at levels that constitute a health threat. However, some regional or countrywide estimates for drinking water exposure exist. In the Bengal delta, 31 million people are exposed to water with $>50\,\mu g\,L^{-1}$ and 50 million to $>10\,\mu g\,L^{-1}$ of As (Chakraborti *et al.*, 2002), which is the WHO, EU and USA limit.

Regulatory agencies have published the MCL for As in drinking water to protect human health. According to the WHO and USEPA guidelines the current recommended limit of As in drinking-water is $10\,\mu g\,L^{-1}$, although this value is designated as provisional because of measurement difficulties and the practical difficulties in removing As from drinking water (USEPA, 2001; WHO, 2001).

1.4.3 *Arsenic species*

Arsenic can be found in two primary forms; organic and inorganic. Organic species of As are mainly found in food, such as shellfish, and include forms as monomethyl arsenic acid (MMAA), dimethyl arsenic acid (DMAA) and arseno-sugars. Inorganic Arsenic occurs in several oxidation states (−3, 0, +3, +5), although in a natural aquatic environment, the existence of As (0) and As (−3) are scarce (Oremland and Stolz, 2003). Groundwater is enriched with inorganic As species where As(V) is the dominant form under oxidizing conditions and exists as $H_2AsO_4^-$, $HAsO_4^{2-}$, AsO_4^{3-} and H_3AsO_4. As(III) is dominant under reducing conditions and exists as $H_2AsO_3^-$ and H_3AsO_3 species (Ng *et al.*, 2004). As(III) is several times more toxic than As(V) (Jain and Ali, 2000) and the concentrations of these species in groundwater are varied depending on the redox conditions of the aquifer organoarsenic compounds are commonly found in surface water that is affected by industrial pollution (Smedley and Kinniburgh, 2002). As(V) is a soft acid and easily

Table 1.1. Aqueous forms of arsenic (adapted from USEPA, 2003).

Arsenite As(III)	H_3AsO_3, $H_2AsO_3^-$, $HAsO_3^{2-}$, AsO_3^{3-}
Arsenate As(V)	H_3AsO_4, $H_2AsO_4^-$, $HAsO_4^{2-}$, AsO_4^{3-}

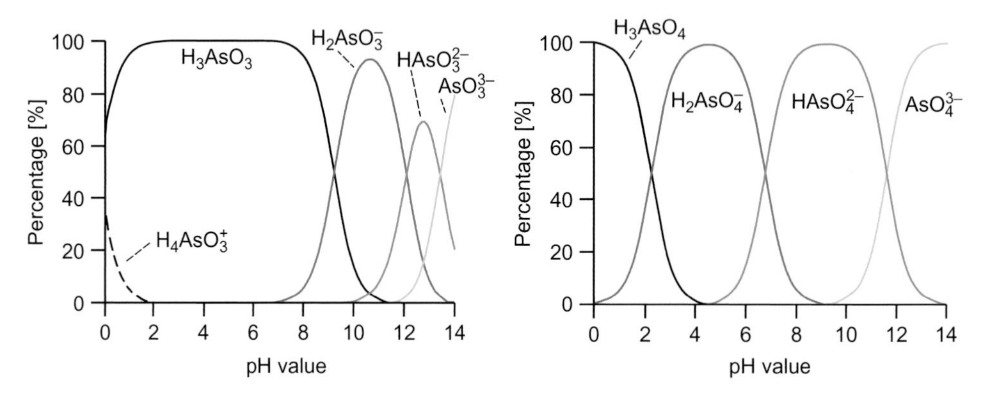

Figure 1.5. Speciation of As(III), (left), and As(V), (right). Total As concentration: 100 μg L^{-1} (adapted from Höll, 2009).

forms complexes with soft bases like sulfide whereas As(III) is a hard acid and prefers to bind with hard bases (e.g., oxides and nitrogen) (Bodek *et al.*, 1998). While As(V) is negatively charged, As(III) is neutral below pH 9 (Smedley and Kinniburgh, 2002) and thus is very difficult to remove from water by conventional technologies.

The formation and concentration change with solution pH of As(III) and As(V) as illustrated in Figure 1.5.

REFERENCES

Alarcon-Herrera, M.T., Bundschuh, J., Nath, B., Nicolli, H., Gutierrez, M., Reyes-Gomez, V.M., Nunez, D., Martin-Dominguez, I.R. & Sracek, O. (2013) Co-occurrence of arsenic and fluoride in groundwater of semi-arid regions in Latin America: genesis, mobility and remediation. *Journal of Hazardous Materials*, 262, 960–969.

Albores, A., Brambila, C., Calderón, A., Del Razo, L., Quintanilla, V. & Manno, M. (2001) Stress proteins induced by arsenic. *Toxicology and Applied Pharmacology*, 177, 132–148.

An, H.K., Park, B.Y. & Kim, D.S. (2001) Crab shell for the removal of heavy metals from aqueous solutions. *Water Research*, 35, 3551–3556.

ATSDR (Agency for Toxic Substances & Disease Registry) *Public Health Statement: arsenic*. US Department of Health and Human Services, Department of Health and Human Services, Public Health Service, Agency for Toxic Substances and Disease Registry. Available from: http://www.atsdr.cdc.gov/ToxProfiles/tp2-c1-b.pdf [accessed April 2015].

Ayotte, J.D., Nielson, M.G. & Robinson, G.R. (1998) Relation of arsenic concentrations in ground water to bedrock lithology in eastern New England. *Geological Society of America Annual Meeting, Abstracts with Programs, St. Louis, MO*. A-58.

Baas Becking, L.G.M., Kaplan, I.R. & Moore, D. (1960) Limits of the natural environment in terms of pH and oxidation-reduction potentials. *The Journal of Geology*, 68, 243–284.

Bates, M.N., Smith, A.H. & Hopenhayn-Rich, C. (1992) Arsenic ingestion and internal cancers: a review. *American Journal of Epidemiology*, 135, 462–475.

Berg, M., Tran, H.C., Nguyen, T.C., Phan, H.V., Schertenleib, R. & Giger, W. (2001) Arsenic contamination of groundwater and drinking water in Vietnam: a human health threat. *Environmental Science & Technology*, 35, 2621–2626.

Bhatnagar, A., Kumar, E. & Sillanpää, M. (2011) Fluoride removal from water by adsorption: a review. *Chemical Engineering Journal*, 171, 811–840.

Bhattacharya, P., Frisbie, S.H., Smith, E., Naidu, R., Jacks, G. & Sarkar, B. (2002a) Arsenic in the environment: a global perspective. In: Sarkar, B. (ed.) *Handbook of heavy metals in the environment*. Marcel Dekker, New York, NY. pp. 145–215.

Bhattacharya, P., Jacks, G., Ahmed, K.M., Khan, A.A. & Routh, J. (2002b) Arsenic in groundwater of the Bengal Delta Plain aquifers in Bangladesh. *Bulletin of Environmental Contamination and Toxicology*, 69, 538–545.

Bhattacharya, P., Classon, M., Bundschuh, J., Sracek, O., Fagerberg, J., Jacks, G., Martin, R.A., Storniolo, A. del R. & Thir, J.M. (2006) Distribution and mobility of arsenic in the río Dulce alluvial aquifers in Santiago del Estero Province, Argentina. *Science of the Total Environment*, 358, 97–120.

Bhattacharya, P., Welch, A.H., Stollenwerk, K.G., McLaughlin, M., Bundschuh, J. & Panaullah, G. (eds.) (2007a) Arsenic in the environment: biology and chemistry. Special Issue. *Science of the Total Environment*, 379 (2–3), 109–265.

Bhattacharya, P., Welch, A.H., Stollenwerk, K.G, McLaughlin, M., Bundschuh, J. & Panaullah, G. (2007b) Arsenic in the environment: biology and chemistry. *Science of the Total Environment*, 379 (2–3), 109–120.

Bhowmick, S., Nath, B., Halder, D., Biswas, A., Majumder, S., Mondal, P., Chakraborty, S., Nriagu, J., Bhattacharya, P., Iglesias, M., Roman-Ross, G., Mazumder, D.G., Bundschuh, J. & Chatterjee, D. (2013) Arsenic mobilization in the aquifers of three physiographic settings of West Bengal, India: understanding geogenic and anthropogenic influences. *Journal of Hazardous Materials*, 262, 915–923.

Birkle, P., Bundschuh, J. & Sracek, O. (2010) Mechanisms of arsenic enrichment in geothermal and petroleum reservoirs fluids in Mexico. *Water Research*, 44 (19), 5605–5617.

Bodek, I., Lyman, W.J., Reehl, W.F. & Rosenblatt, D.H. (1998) *Environmental inorganic chemistry: properties, processes and estimation methods*. Pergamon Press, New York, NY.

Borgoño, J.M. & Greiber, R. (1971) Epidemiological study of arsenicism in the city of Antofagasta. *Trace Substances in Environmental Health*, 5, 13–24.

Borja, A., Calderón, J., Díaz Barriga, F., Goleen, A., Jiménez, C., Navarro, M., Rodríguez, L. & Santos, D. (2001) Exposure to arsenic and lead and neuropsychological development in Mexican children. *Environmental Research*, Section A, 85, 69–76.

Boudette, E.L., Canney, F.C., Cotton, J.E., Davis, R.I., Ficklin, W.H. & Motooka, J.M. (1985) High levels of arsenic in the groundwater of southeastern New Hampshire: a geochemical reconnaissance. USGS Open-File Report 85-202.

Bugaisa, S.L. (1971) Significance of fluorine in Tanzania drinking water. *Proceedings Conference Rural Water Supply in East Africa, 5–8 April 1971, Dar-Es-Salaam, Tanzania*.

Bundschuh, J. & Prakash Maity, J. (2015) Geothermal arsenic: occurrence, mobility and environmental implications. *Renewable & Sustainable Energy Reviews*, 42, 1214–1222.

Bundschuh, J., Bonorino, G., Viero, A.P., Albouy, R. & Fuertes, A. (2000) Arsenic and other trace elements in sedimentary aquifers in the Chaco-Pampean Plain, Argentina: origin, distribution, speciation, social and economic consequences. In: Bhattacharya, P. & Welch, A.H. (eds.) Arsenic in groundwater of sedimentary aquifers. *Pre-Congress Workshop, International Geological Congress, 3–5 August 2000, Rio de Janeiro, Brazil, 2000*. pp. 27–32.

Bundschuh, J., Farías, B., Martin, R., Storniolo, A., Bhattacharya, P., Cortes, J., Bonorino, G. & Albouy, R. (2004) Groundwater arsenic in the Chaco-Pampean Plain, Argentina: case study from Robles County, Santiago del Estero Province. *Applied Geochemistry*, 19, 231–243.

Bundschuh, J., Pérez Carrera, A. & Litter, M.I. (eds.) (2008a) *Distribución del arsénico en las regiones Ibérica e Iberoamericana*. Editorial Programa Iberoamericano de Ciencia y Tecnologia para el Desarrollo, Buenos Aires, Argentina.

Bundschuh, J., Giménez-Forcada, E., Guérèquiz, R., Pérez-Carrera, A., Garcia, M.E., Mello, J. & Deschamps, E. (2008b) Fuentes geogénicas de arsénico y su liberación al medio ambiente. In: Bundschuh, J., Pérez-Carrera, A. & Litter, M.I. (eds.) *Distribución del arsénico en las regiones Ibérica e Iberoamericana*. Editorial Programa Iberoamericano de Ciencia y Tecnologia para el Desarrollo, Buenos Aires, Argentina. pp. 33–47.

Bundschuh, J., Nicolli, H.B., Blanco, M. del C., Blarasin, M., Farías, S.S., Cumbal, L., Cornejo, L., Acarapi, J., Lienqueo, H., Arenas, M., Guérèquiz, R., Bhattacharya, P., García, M.E., Quintanilla, J., Deschamps, E., Viola, Z., Castro de Esparza, M.L., Rodríguez, J., Pérez-Carrera, A. & Fernández Cirelli, A. (2008c)

Distribución de arsénico en la región sudamericana. In: Bundschuh, J., Pérez-Carrera, A. & Litter, M.I. (eds.) *Distribución del arsénico en las regiones Ibérica e Iberoamericana*. Editorial Programa Iberoamericano de Ciencia y Tecnologia para el Desarrollo, Buenos Aires, Argentina. pp. 137–186.
Bundschuh, J., Armienta, M.A., Birkle, P., Bhattacharya, P., Matschullat, J. & Mukherjee, A.B. (eds.) (2009a) *Natural arsenic in groundwater of Latin America*. CRC Press, Boca Raton, FL.
Bundschuh, J., García, M.E., Birkle, P., Cumbal, L.H., Bhattacharya, P. & Matschullat, J. (2009b) Occurrence, health effects and remediation of arsenic in groundwaters of Latin America. In: Bundschuh, J., Armienta, M.A., Birkle, P., Bhattacharya, P., Matschullat, J. & Mukherjee, A.B. (eds.) *Natural arsenic in groundwater of Latin America*. CRC Press, Boca Raton, FL. pp. 3–15.
Bundschuh, J., Bhattacharya, P., von Brömssen, M. Jakariya, M., Jacks, G., Thunvik, R. & Litter, M.I. (2009c) Arsenic-safe aquifers as a socially acceptable source of safe drinking water – what can rural Latin America learn from Bangladesh experiences? In: Bundschuh, J., Armienta, M.A., Birkle, P., Bhattacharya, P., Matschullat, J. & Mukherjee, A.B. (eds.) *Natural arsenic in groundwater of Latin America*. CRC Press, Boca Raton, FL. pp. 687–697.
Bundschuh, J., Litter, M.I. & Bhattacharya, P. (2010) Safe water production by targeting arsenic-safe aquifers. *Environmental Geochemistry and Health*, 32 (4), 307–315.
Bundschuh, J., Litter, M.I., Parvez, F., Román-Ross, G., Nicolli, H.B., Jean, J.-S., Liu, C.-W., López, D., Armienta, M.A., Guilherme, L.R.G., Cuevas, A.H., Cornejo, L., Cumbal, L. & Toujaguez, R. (2012a) One century of arsenic exposure in Latin America: a review of history and occurrence from 14 countries. *Science of the Total Environment*, 429, 2–35.
Bundschuh, J., Nath, B., Bhattacharya, P., Liu, C.-W., Armienta, M.A., Moreno López, M.V., Lopez, D.L., Jean, J.-S., Cornejo, L., Fagundes, L., Macedo, L. & Tenuta Filho, A. (2012b) Arsenic in the human food chain: the Latin American perspective. *Science of the Total Environment*, 429, 92–106.
Bundschuh, J., Prakash Maity, J., Nath, B., Baba, A., Gunduz, O., Kulp, T.T., Jean, J.-S., Kar, S., Tseng, Y.-J., Bhattacharya, P. & Chen, C.Y. (2013) Naturally occurring arsenic in terrestrial geothermal systems of western Anatolia, Turkey: potential role in contamination of freshwater resources. *Journal of Hazardous Materials*, 262, 951–959.
Carter, J.M., Sando, S.K., Hayes, T.S. & Hammond, R.H. (1998) Source, occurrence, and extent of arsenic contamination in the grass mountain area of the Rosebud Indian Reservation, South Dakota. USGS Water Resources Investigation Report 97-4286.
Chakraborti, D., Rahman, M.M., Pau, K., Chowdhury, U.K., Sengupta, M.K., Lodh, D., Chanda, C.R., Saha, K.C. & Mukherjee, S.C. (2002) Arsenic calamity in the Indian subcontinent: what lessons have been learned? *Talanta*, 58, 3–22.
Chandrasekharam, D. & Bundschuh, J. (2008) *Low-enthalpy geothermal resources for power generation*. CRC Press, Boca Raton, FL.
Chatterjee, D., Halder, D., Ashis Biswas, A., Bhattacharya, P., Mazumder, S., Bhowmick, S., Mukherjee-Goswami, A., Saha, D., Maity, P.B., Chatterjee, D., Nath, B., Mukherjee, A. & Bundschuh, J. (2010) Assessment of arsenic exposure from groundwater and rice in Bengal Delta Region, West Bengal, India. *Water Research*, 44 (19), 5803–5812.
Cheng, T. Barnett, M.O., Roden, E.E. & Zhuang, J. (2004) Effects of phosphate on U(VI) adsorption to goethite-coated sand. *Environmental Science & Technology*, 38 (22), 6059–6065.
Chinoy, N.J. (1991) Effects of fluoride on physiology of animals and human beings. *Indian Journal of Environment and Toxicology*, 1, 17–32.
Cullen, W.R. & Reimer, K.J. (1989) Arsenic speciation in the environment. *Chemical Reviews*, 89, 713–764.
Dangic, A. & Dangic, J. (2007) Arsenic in the soil environment of central Balkan Peninsula, southeastern Europe: occurrence, geochemistry, and impacts. In: Bhattacharya, P., Mukherjee, A.B., Bundschuh, J., Zevenhoven, R. & Loeppert, R.H. (eds.) *Arsenic in soil and groundwater environment*. Elsevier, Amsterdam, The Netherlands. pp. 207–236.
Davies, J., Davis, R., Frank, D., Frost, F., Garland, D., Milham, S., Pierson, R.S., Raasina, R.S., Safioles, S. & Woodruff, L. (1991) *Seasonal study of arsenic in ground water: Snohomish County, Washington*. Snohomish Health District and Washington State Department of Health.
Del Razo, L., Gonsebatt, M., Gutiérrez, M. & Ramírez, P. (2000) Arsenite induces DNA-protein crosslink and cytokeratin expression in the WRL-68 human hepatic cell line. *Carcinogenesis*, 21 (4), 701–706.
Del Razo, L., De Vizcaya, R., Izquierdo, V., Sanchez, P. & Soto, C. (2005) Diabetogenic effects and pancreatic oxidative damage in rats subcronically exposed to arsenite. *Toxicology Letters*, 160, 135–142.
Endo, G., Fukushima, S., Kinoshita, A., Kuroda, K., Morimura, K., Salim, E., Shen, J., Wanibuchi, H., Wei, M. & Yoshida, K. (2003) Understanding arsenic carcinogenicity by the use of animal models. *Toxicology and Applied Pharmacology*, 198, 366–376.

Favre-Réguillon, A., Lebuzit, G., Foos, J., Guy, A., Draye M. & Lemaire, M. (2003) Selective concentration of uranium from seawater by nanofiltration. *Industrial & Engineering Chemistry Research*, 42, 5900–5904.

Favre-Réguillon, A., Lebuzit, G., Foos, J., Guy, A., Sorin, A., Lemaire, M. & Draye, M. (2005) Selective rejection of dissolved uranium carbonate from seawater using crossflow filtration technology. *Separation Science and Technology*, 40 (1–3), 623–631.

Fawell, J., Bailey, K., Chilton, J., Dahi, E., Fewtrell, L. & Magara, Y. (2006) Fluoride in drinking-water. World Health Organization (WHO): Guidelines for drinking water quality: first addendum to third edition. Volume 1: Recommendations. World Health Organisation, Geneva, Switzerland. IWA Publishing, London, UK.

Feldman, P.R. & Rosenboom, J.W. (2001) Cambodia drinking water quality assessment. WHO in cooperation with Cambodian Ministry of Rural Development and the Ministry of Industry, Mines, and Energy. Phnom Penh, Cambodia.

Ficklin, W.H., Frank, D.G., Briggs, P.K. & Tucker, R.E. (1989) Analytical results for water, soil, and rocks collected near the vicinity of Granite Falls Washington as part of an arsenic-in-groundwater study. USGS Open-File Report 89-148.

Fontaine, J.A. (1994) Regulating arsenic in Nevada drinking water supplies: past problem, future challenges. In: Chappell, W.R., Abernathy, C.O. & Cothern, R.L. (eds.) Arsenic exposure and health effects. *Science and Technology Letters*, 285–288.

Frisbie, S.H., Mitchell, E.J. & Sarkar (2013) World Health Organization increases its drinking-water guideline for uranium. *Environmental Science: Processes & Impacts*, 15 (10), 1817–1823.

Fujii, R. & Swain, W.C. (1995) Areal distribution of trace elements, salinity, and major ions in shallow ground water, Tulare Basin, southern San Joaquin Valley, California. USGS Water Resources Investigation Report 95-4048.

Gbadebo, A.M. (2005) Occurence and fate of arsenic in the hydrogeological systems of Nigeria. *Geological Society of America Abstracts with Programs*, 37 (7), 375.

Goldblatt, E.L., Van Denburgh, S.A. & Marsland, R.A. (1963) The unusual and widespread occurrence of arsenic in well waters of Lane Country, Oregon. Lane County Health Department Report, Eugene, OR.

Graham, G.S., Kesler, S.E. & Van Loon, J.C. (1975) Fluorine in ground water as a guide to Pb-Zn-Ba-F mineralization. *Economic Geology*, 70, 396–398.

Grenthe, I., Fuger, J., Konings, R.J.M., Lemire, R.J., Muller, A.B., Nguyen-Trung, C. & Wanner, H. (1992) *Chemical thermodynamics of uranium*. North-Holland, Amsterdam, The Netherlands.

Harrison, P.T.C. (2005) Fluoride in water: a UK perspective. *Journal of Fluorine Chemistry*, 126, 1448–1456.

Hasan, M.A., Ahmed, K.M., Sracek, O., Bhattacharya, P., von Brömssen, M., Broms, S., Fogelstrom, J., Mazumder, M.L. & Jacks, G. (2007) Arsenic in shallow groundwater of Bangladesh: investigations from three different physiographic settings. *Hydrogeology Journal*, 15 (8), 1507–1522.

Hoinkis, J., Valero-Freitag, S., Martin, Caporgno, P. & Paetzold, C. (2011) Removal of nitrate and fluoride by nanofiltration—a comparative study. *Desalination and Water Treatment*, 30, 278–288.

Höll, W. (2009) Mechanisms of arsenic removal from water. In: Kabay, N., Bundschuh, J., Hendry, B., Bryjak, M., Yoshizuka, K., Bhattacharya, P. & Anaç, S. (eds.) *Global arsenic problem and challenges for safe water production*. CRC Press, Boca Raton, FL. pp. 49–58.

Jacks, G. & Battacharya, P. (2009) Drinking water from groundwater sources—a global perspective. In: Forare, J. (ed.) Drinking water: sources, sanitation and safeguarding. The Swedish Research Council Formas, Stockholm, Sweden. pp. 21–33.

Jain, C.K. & Ali, I. (2010) Arsenic occurrence, toxicity and speciation techniques. *Water Research*, 34 (17), 4304–4312.

Jean, J.-S, Bundschuh, J., Chen, C.-J., Guo, H.-R., Liu, C.-W., Lin, T.-F. & Chen, Y.-H. (2010) *The Taiwan crisis: a showcase of the global arsenic problem*. CRC Press, Boca Raton, FL.

Kanno, M. (1984) Present status of study on extraction of uranium from sea water. *Journal of Nuclear Science and Technology*, 21, 1–9.

Kawai, T., Saito, K., Sugita, K., Katakai, A., Seko, N., Sugo, T., Kanno, J.I. & Kawakami, T. (2008) Comparison of amidoxime adsorbents prepared by cografting methacrylic acid and 2-hydroxyethyl methacrylate with acrylonitrile onto polyethylene. *Industrial & Engineering Chemistry Research*, 39, 2910–2915.

Khatibikamala, V., Torabiana, A., Janpoora, F. & Hoshyaripourb, G. (2010) Fluoride removal from industrial wastewater using electrocoagulation and its adsorption kinetics. *Journal of Hazardous Materials*, 179, 276–280.

Kilham, P. & Hecky, R.E. (1973) Fluoride: geochemical and ecological significance in East African waters and sediments. *Limnology and Oceanography*, 18, 932–945.

Kirk, T. & Sarfaraz, A. (2003) Oxidative stress as a possible mode of action for arsenic carcinogenesis. *Toxicology Letters*, 137, 3–13.
Kobuke, Y., Tabushi, I., Aoki, T., Kamaishi, T. & Hagiwara, I. (1988) Composite fiber adsorbent for rapid uptake of uranyl from seawater. *Industrial & Engineering Chemistry Research*, 27, 1461–1466.
Kumar, J.R., Kim, J.-S., Lee, J.-Y. & Yoon, H.-S. (2011) A brief review on solvent extraction of uranium from acidic solutions. *Separation & Purification Reviews*, 40, 77–125.
Langmuir, D. (1997) *Aqueous environmental geochemistry*. Prentice Hall, Upper Saddle River, NJ.
Lazo, P., Cullaj, A., Arapi, A. & Deda, T. (2007) Arsenic in soil environments in Albania. In: Bhattacharya, P., Mukherjee, A.B., Bundschuh, J., Zevenhoven, R. & Loeppert, R.H. (eds.) *Arsenic in soil and groundwater environment*. Elsevier, Amsterdam, The Netherlands. pp. 237–256.
Liu, C., Bai, R. & Ly, Q.S. (2008) Selective removal of copper and lead ions by diethylenetriamine-functionalized adsorbent: behaviours and mechanisms. *Water Research*, 42, 1511–1522.
Liu, C.-C., Prakash Maity, J., Jean, J.-S., Sracek, O., Kar, S., Li, Z., Bundschuh, J., Chen, C.-Y. & Lu, H.-Y. (2011) Biogeochemical interactions among the arsenic, iron, humic substances, and bacterial activities in mud volcanoes in southern Taiwan. *Journal of Environmental Sciences Health* A, 46 (11), 1218–1230.
Liu, C.-C., Kar, S., Jean, J.-S., Wang, C.-H., Lee, Y.-C., Sracek, O., Li, Z., Bundschuh, J., Yang, H.-J. & Chen, C.-Y. (2013) Linking geochemical processes in mud volcanoes with arsenic mobilization driven by organic matter. *Journal of Hazardous Materials*, 262, 980–988.
López, D.L., Bundschuh, J., Birkle, P., Armienta, M.A., Cumbal, L., Sracek, O., Cornejo, L. & Ormachea, M. (2012) Arsenic in volcanic geothermal fluids of Latin America. *Science of the Total Environment*, 429, 57–75.
Mahramanlioglu, M., Kizilcikli, I. & Bicer, I.O. (2002) Adsorption of fluoride from aqueous solution by acid treated spent bleaching earth. *Journal of Fluorine Chemistry*, 115, 41–47.
Matisoff, G., Khourey, C.J., Hall, J.F., Varnes, A.W. & Strain, W.H. (1982) The nature and source of arsenic in northeastern Ohio ground water. *Ground Water*, 20 (4), 446–456.
McClintock, T.R., Chen, Y., Bundschuh, J., Oliver, J.T., Navoni, J., Olmos, V., Villaamil, E., Habibul Ahsan, L.H. & Parvez, F. (2012) Arsenic exposure in Latin America: biomarkers, risk assessments and related health effects. *Science of the Total Environment*, 429, 76–91.
Mohapatra, M., Anand, S., Mishra, B.K., Giles, D.E. & Singh, P. (2009) Review of fluoride removal from drinking water. *Journal of Environmental Management*, 91 (1), 67–77.
Mondal, P., Bhowmick, S., Chatterjee, D., Figoli, A. & Van der Bruggen, B. (2013) Remediation of inorganic arsenic in groundwater for safe water supply: a critical assessment of technological solutions. *Chemosphere*, 92, 157–170.
Mukherjee, A., Fryar, A.E. & Howell, P. (2007) Regional hydrostratigraphy and groundwater flow modeling of the arsenic contaminated aquifers of the western Bengal basin, West Bengal, India. *Hydrogeology Journal*, 15, 1397–1418.
Mulligan, C.N., Yong, R.N. & Gibbs, B.F. (2001) Heavy metal removal from sediments by biosurfactants. *Journal of Hazardous Materials*, 85, 111–125.
Nadakavukaren, J.J., Ingermann, R.L. & Jeddeloh, G. (1984) Seasonal variation of arsenic concentration in well water in Lane County, Oregon. *Bulletin of Environmental Contamination and Toxicology*, 33 (3), 264–269.
Nanyaro, J.T., Aswathanarayana, U. & Mungure, J.S. (1984) A geochemical model for the abnormal fluoride concentrations in waters in parts of northern Tanzania. *Journal of African Earth Sciences*, 2, 129–140.
Ng, K.S., Ujang, Z. & Le-Clech, P. (2004) Arsenic removal technologies for drinking water treatment. *Reviews in Environmental Science and Bio/Technology*, 3 (1), 43–53.
Nickson, R.T., McArthur, J.M., Shrestha, B., Kyaw-Myint, T.O. & Lowry, D. (2005) Arsenic and other drinking water quality issues, Muzaffargarh District, Pakistan. *Applied Geochemistry*, 20, 55–68.
Nicolli, H.B., Bundschuh, J., García, J.W., Falcón, C.M. & Jean, J.-S. (2010) Sources and controls for the mobility of arsenic in oxidizing groundwaters from loess-type sediments in arid/semi-arid dry climates – evidence from the Chaco-Pampean plain (Argentina). *Water Research*, 44 (19), 5589–5604.
Nicolli, H.B., Bundschuh, J., Blanco, M. del C., Tujchneider, O.C., Panarello, H.O., Dapeña, C. & Rusansky, H.E. (2012) Arsenic and associated trace-elements in groundwater from the Chaco-Pampean plain, Argentina: results from 100 years of research. *Science of the Total Environment*, 429, 36–56.
OECD (2000) Uranium 1999: resources, production and demand. OECD Nuclear Energy Agency and the International Atomic Energy Agency, Paris, France.
OECD (2009) Uranium 2009: resources, production and demand. Publication 6891, OECD Nuclear Energy Agency, Paris, France, 2010. Available from: https://www.oecd-nea.org/ndd/pubs/2010/6891-uranium-2009.pdf [accessed April 2015].

OEHHA (2001) Public health goal for chemicals in drinking water: uranium. Prepared by the Office of Environmental Health Hazard Assessment, California Environmental Protection Agency, CA, USA, August 2001.

Oliver, I.W., Graham, M.C., MacKenzie, A.B., Ellam, R.M. & Farmer, J.G. (2008) Distribution and partitioning of depleted uranium (DU) in soils at weapons test ranges—investigations combining the BCR extraction scheme and isotopic analysis. *Chemosphere*, 72 (6), 932–939.

Ono, F.B., Guilherme, L.R.G., Silva Penido, E., Santos Carvalho, G., Hale, B., Toujaguez, R. & Bundschuh, J. (2012) Arsenic bioaccessibility in a gold mining area: a health risk assessment for children. *Environmental Geochemistry and Health*, 34, 457–465.

Oremland, S. & Stolz, J.F. (2003) The ecology of arsenic. *Science*, 300 (5621), 939–944.

Orloff, K.G., Mistry, K., Charp, P., Metcalf, S., Marino, R., Shelly, T., Melaro, E., Donohoe, A.M. & Jones, R.L. (2004) Human exposure to uranium in groundwater. *Environmental Research*, 94 (3), 319–326.

Ormachea Muñoz, M., Wern, H., Johnsson, F., Bhattacharya, P., Sracek, O., Thunvik, R., Quintanilla, J. & Bundschuh, J. (2013) Geogenic arsenic and other trace elements in the shallow hydrogeologic system of southern Poopó Basin, Bolivian Altiplano. *Journal of Hazardous Materials*, 262, 924–940.

Ormachea Muñoz, M., Bhattacharya, P., Sracek, O., Ramos Ramos, O., Quintanilla Aguirre, J., Bundschuh, J. & Prakash Maity, J. (2015) Arsenic and other trace elements in thermal springs and in cold waters from drinking water wells on the Bolivian Altiplano. *Journal of South American Earth Sciences*, 60, 10–20.

Peters, S.C. (2008) Arsenic in groundwaters in the northern Appalachian mountain belt: a review of patterns and processes. *Journal of Contaminant Hydrology*, 99 (1–4), 8–21.

Peters, S.C., Blum, J.D., Klaue, B. & Karagas, M.R. (1999) Arsenic occurrrence in New Hampshire ground water. *Environmental Science & Technology*, 33 (9), 1328–1333.

Prakash Maity, J., Nath, B., Chen, C.-Y., Bhattacharya, P., Sracek, O., Bundschuh, J., Thunvik, R., Kar, S., Chatterjee, D. & Jean, J.-S. (2011a) Arsenic enrichment in groundwaters of Bengal Basin (India, Bangladesh) and Chianan Plain (Taiwan): comparison of hydrochemical characteristics and mobility constraints. *Journal of Environmental Sciences Health* A, 46 (11), 1163–1176.

Prakash Maity, J., Liu, C.-C., Nath, B., Bundschuh, J., Kar, S., Jean, J.-S., Bhattacharya, P., Liu, J.-H., Atla, S.-B. & Chen, C.-Y. (2011b) Biogeochemical characteristics of Kuan-Tzu-Ling, Chung-Lun and Bao-Lai hot springs in southern Taiwan. *Journal of Environmental Sciences Health* A, 46 (11), 1207–1217.

Raff, O. & Wilken, R.D. (1999) Removal of dissolved uranium by nanofiltration. *Desalination*, 122, 147–150.

Ramos Ramos, O.E., Cáceres, L.F., Ormachea Munñoz, M.R., Bhattacharya, P., Quino, I., Quintanilla, J., Sracek, O., Thunvik, R., Bundschuh, J. & García, M.E. (2012) Sources and behavior of arsenic and trace elements in groundwater and surface water in the Poopó Lake Basin, Bolivian Altiplano. *Environmental Earth Sciences*, 66 (3), 793–807.

Ramos Ramos, O.E., Rötting, T.S., French, M., Sracek, O., Bundschuh, J., Quintanilla, J. & Bhattacharya, P. (2014) Geochemical processes controlling mobilization of arsenic and trace elements in shallow aquifers and surface waters in the Antequera and Poopó mining regions. *Journal of Hydrology*, 518, 421–433.

Ravenscroft, P. (2007) Predicting the global extent of arsenic pollution of groundwater and its potential impact on human health. Report prepared for UNICEF, New York, NY.

Raychowdhury, N., Mukherjee, A., Bhattacharya, P., Johannesson, K., Bundschuh, J., Bejarano Sifuentes, G., Nordberg, G., Martin, R.A. & Storniolo, A. del R. (2013) Provenance and fate of arsenic and other solutes in the Chaco-Pampean Plain of the Andean foreland, Argentina: from perspectives of hydrogeochemical modeling and regional tectonic setup. *Journal of Hydrology*, 518 (Part C), 300–316.

Reimann, C. & Banks, D. (2004) Setting action levels for drinking water: are we protecting our health or our economy (or our backs!)? *Science of the Total Environment*, 332, 13–21.

Reza, A.H.M.S., Jean, J.-S., Lee, M.-K., Luo, S.-D., Bundschuh, J., Li, H.-C., Yang, H.-J. & Liu, C.-C. (2011a) Interrelationship of TOC, As, Fe, Mn, Al and Si in shallow alluvial aquifers in Chapai-Nawabganj, northwestern Bangladesh: implication for potential source of organic carbon. *Environmental Earth Sciences*, 63 (5), 955–967.

Reza, A.H.M.S., Jean, J.-S., Yang, H.-J., Lee, M.-K., Hsu, H.-F., Liu, C.-C., Lee, Y.-C., Bundschuh, J., Lin, K.-H. & Lee, C.-Y. (2011b) A comparative study on arsenic and humic substances in alluvial aquifers of Bengal delta plain (NW Bangladesh), Chianan plain (SW Taiwan) and Lanyang plain (NE Taiwan): implication of arsenic mobilization mechanisms. *Environmental Geochemistry and Health*, 33 (3), 235–258.

Roberson, C.E. & Barnes, R.B. (1978) Stability of fluoride complex with silica and its distribution in natural water systems. *Chemical Geology*, 21, 239–256.

Robertson, F.N. (1984) Solubility controls of fluorine, barium and chromium in ground water in alluvial basins of Arizona. Geological Survey Tucson, Water Resources Division. In: *First Canadian/American Conference on Hydrogeology: Practical Applications of ground water geochemistry, 22–26 June 1984, Banff, Alberta, Canada*. pp. 96–102.

Rossiter, H.M.A., Graham, M.C. & Schäfer, A.I. (2010) Impact of speciation on behaviour of uranium in a solar powered membrane system for treatment of brackish groundwater. *Separation and Purification Technology*, 71 (1), 89–96.

Rossman, T. (2003) Mechanism of arsenic carcinogenesis: an integrated approach: fundamental and molecular mechanisms of mutagenesis. *Mutation Research*, 533, 37–65.

Schwochau, K. (1984) Extraction of metals from sea water. *Topics in Current Chemistry*, 124, 91–133.

Shen, F., Chen, X., Gao, P. & Chen, G. (2003) Electrochemical removal of fluoride ions from industrial wastewater. *Chemical Engineering Science*, 58, 987–993.

Sheppard, S.C., Sheppard, M.I., Gallerand, M.O. & Sanipelli, B. (2005) Derivation of ecotoxicity thresholds for uranium. *Journal of Environmental Radioactivity*, 79, 55–83.

Smedley, P.L. (1996) Arsenic in rural groundwater in Ghana. *Journal of African Earth Sciences*, 22 (4), 459–470.

Smedley, P.L. (2006) Sources and distribution of arsenic in groundwater and aquifers. In: *Proceedings of the Symposium Arsenic in Groundwater – A World Problem, 29 November 2006b, Utrecht, The Netherlands*. pp. 4–32.

Smedley, P.L. & Kinniburgh, D.G. (2002) A review of the source, behaviour and distribution of arsenic in natural waters. *Applied Geochemistry*, 17 (5), 517–568.

Smedley, P.L., Kinniburgh, D.G., Macdonald, D.M.J., Nicolli, H.B., Barros, A.J., Tullio, J.O., Pearce, J.M. & Alonso, M.S. (2005) Arsenic associations in sediments from the loess aquifer of La Pampa, Argentina. *Applied Geochemistry*, 20, 989–1016.

Smedley, P.L., Smith, B., Abesser, C. & Lapworth, D. (2006) Uranium occurrence and behaviour in British groundwater. Groundwater Programme, Commissioned Report CR/06/050N, Nottingham British Geological Survey, Keyworth, UK.

Smedley, P.L., Nicolli, H.B., Macdonald, D.M.J. & Kinniburgh, D.G. (2009) Arsenic in groundwater and sediments from La Pampa province, Argentina. In: Bundschuh, J., Armienta, M.A., Birkle, P., Bhattacharya, P., Matschullat, J. & Mukherjee, A.B. (eds.) *Natural arsenic in groundwater of Latin America*. CRC Press, Boca Raton, FL. pp. 35–45.

Sujana, M.G., Thakur, R.S. & Rao, S.B. (1998) Removal of fluoride from aqueous solution by using alum sludge. *Journal of Colloid and Interface Science*, 206, 94–101.

Swartz, R.J. (1995) *A study of the occurrence of arsenic on the Kern Fan Element of the Kern Water Bank, Southern San Joaquin Valley, California*. MSc Thesis. Department of Geology, California State University, Bakersfield, CA.

Swartz, R.J., Thyne, G.D. & Gillespie, J.M. (1996) Dissolved arsenic in the Kern Fan San Joaquin Valley, California: naturally occurring or anthropogenic. *Environmental & Engineering Geoscience*, 3 (3), 143–153.

The Royal Society (2001) *The health hazards of depleted uranium munitions*: Part I. London, UK.

Toujaguez, R., Ono, F.B., Martins, V., Bundschuh, J., Cabrera, P.P., Blanco, A.V. & Guilherme, L.R.G. (2013) Arsenic bioaccessibility in gold mine tailings of Delita, Cuba. *Journal of Hazardous Materials*, 262, 1004–1013.

Toyoda, A. & Taira, T. (2000) A new method for treating fluorine wastewater to reduce sludge and running costs. *IEEE Transactions on Semiconductor Manufacturing*, 13, 305–309.

UN (2013) Press release embargoed until 13 June 2013, 11:00 A.M., EDT.

UNICEF (1999) State of the art report on the extent of fluoride in drinking water and the resulting endemicity in India. Report by Fluorosis Research & Rural Development Foundation of UNICEF, New Delhi, India.

USEPA (2001) National primary drinking water regulations; arsenic and clarifications to compliance and new source contaminants monitoring; final rule. *Federal Register*, 66, 6076–7066.

USEPA (2003) Arsenic treatment technology evaluation handbook for small systems. H.A.E.C. Division, Washington, DC.

Vilar, V.J.P., Botelho, C.M.S. & Boaventura, R.A.R. (2007) Copper desorption from Gelidium algal biomass. *Water Research*, 41, 1569–1579.

Wasserman, G., Liu, X., Parvez, F., Ahsan, H., Factor-Litvak, P., van Geen, A., Slavkobich, V., Lolacono, N., Cheng, Z., Hussain, I., Momotaj, H. & Graziano, J. (2004) Water arsenic exposure and children's intellectual function in Araihazar Bangladesh. *Environmental Health Perspectives*, 112 (13), 1329–1333.

Welch, A.H. & Lico, M.S. (1998) Factors controlling As and U in shallow ground water, southern Carson Desert, Nevada. *Applied Geochemistry*, 13 (4), 521–539.
Welch, A.H., Westjohn, D.B., Helsel, D.R. & Wanty, R.B. (2000) Arsenic in ground water of the United States: occurrence and geochemistry. *Ground Water*, 38, 589–604.
Westjohn, D.B., Kolker, A., Cannon, W.F. & Sibley, D.F. (1998) Arsenic in ground water in the "thumb area" of Michigan. In: *The Mississippian Marshall Sandstone Revisited, Michigan: Its Geology and Geologic Resources: 5th Symposium, Michigan Department of Environmental Quality, East Lansing, MI.* pp. 24–25.
WHO (2001) Guidelines for drinking-water quality: arsenic in drinking water. Fact Sheet No. 210. Geneva, Switzerland, World Health Organization.
WHO (2004) Uranium in drinking-water, background document for development of WHO guidelines for drinking-water quality. World Health Organization, Geneva, Switzerland.
WHO (2006) Fluoride in drinking-water. World Health Organisation, Geneva, Switzerland. Available from: http://www.who.int/water_sanitation_health/publications/fluoride_drinking_water_full.pdf [accessed January 2014].
WHO (2008) Guidelines for drinking water quality. Volume 1, Recommendations. 3rd edition. World Health Organization, Geneva, Switzerland.
WRUD (2001) Preliminary study on arsenic contamination in selected areas of Myanmar. Water Resources Utilization Department (WRUD) in the Ministry of Agriculture and Irrigation of Myanmar, Rangun, Myanmar.
Zhou, P. & Gu, B. (2005) Extraction of oxidized and reduced forms of uranium from contaminated soils: effects of carbonate concentration and pH. *Environmental Science & Technology*, 39 (12), 4435–4440.
Zuena, A.J. & Keane, P.E. (1985) Arsenic contamination of private potable wells. *EPA National Conference on Environmental Engineering Proceedings, 717–725. Boston, MA*. USEPA.

Part II
Traditional membrane processes

CHAPTER 2

Arsenic removal by low pressure-driven membrane operations

Alfredo Cassano

2.1 INTRODUCTION

Water sources are often contaminated by micro-pollutants coming from rainfall, sewage, landfill leachate and industrial wastewaters. These micro-pollutants include micro-organisms, dispersed substances, organic compounds, such as natural organic matter (NOM), and inorganic substances. Among different inorganic compounds arsenic (As), uranium (U) and fluoride (F^-) have been found in natural water sources and wastewaters at concentrations potentially dangerous for human health (Bodzek *et al.*, 2011).

In this chapter basic principles of microfiltration (MF) and ultrafiltration (UF) processes and selected applications within the treatment of drinking water for As removal are illustrated and discussed highlighting significant advantages which can be achieved through their integration with adsorption and coagulation/flocculation technologies. These principles can be extended in good approximation to the removal of F^- and U from drinking water.

Conventional methods of As removal from drinking water include oxidation/precipitation, coagulation/flocculation, adsorption, ion-exchange and membrane technologies. Similar approaches can be used also for defluoridation (Meenakshi and Maheshwari, 2006) and removal of U from drinking water (Katsoyiannis and Zouboulis, 2013) although some of these methods have been tested at laboratory or pilot scale only.

Oxidation methods are relatively simple and assure oxidation of other impurities and microbial degradation with minimum residual mass; as drawbacks they mainly remove As(V) and require an efficient control of pH (Zaw and Emett, 2002).

Coagulation/flocculation methodologies are based on the use of commercially available chemicals such as aluminum, ferric sulfate, ferric chloride, slaked or hydrated lime, ferric hydroxide and polyaluminum chloride (Meng *et al.*, 2001; Ng *et al.*, 2004). They are effective over a wider range of pH, simple in operation and characterized by low capital costs. Some disadvantages are in terms of the production of toxic sludge, low removal of As, release of taste and odor compounds due to chlorination, floc disposal and post-treatment (Song *et al.*, 2006; Wickramasinghe *et al.*, 2004).

Adsorption methods based on the use of activated carbon, activated alumina and ion exchanger resins are promising processes for As removal because of the low cost, high efficiency, removal capability of both As(III) and As(V), independence on pH (Kim and Benjamin, 2004; Mohan and Pittman, 2007). The main disadvantages of adsorption methods are the requirement for multiple chemical treatments, high running/capital costs, pre- and/or post-treatments of drinking water, disposal of both spent media and wastewaters produced during regeneration/cleaning of columns.

The use of pressure-driven membrane processes, including microfiltration (MF), ultrafiltration (UF), nanofiltration (NF) and reverse osmosis (RO), for the removal of As from drinking water has been reviewed by Shih (2005) and Uddin *et al.* (2007a). These processes are characterized by low consumption of energy, no requirement for chemical substances to be added, an easy way to increase the capacity (modular system), separation in the continuous mode, the possibility of integration with other unit processes (hybrid processes), and separation carried out in mild environment conditions (Drioli *et al.*, 1999). In particular, the use of NF and RO membranes for the removal of As from drinking water has been largely investigated (Cakmakci *et al.*, 2009;

Figoli *et al.*, 2010; Geucke *et al.*, 2006; Johnston *et al.*, 2001; Kang *et al.*, 2000; Kim *et al.*, 2006; Košutić *et al.*, 2005; Ning, 2002; Oh *et al.*, 2000; Saitúa *et al.*, 2005; Sato *et al.*, 2002; Uddin *et al.*, 2007b; Urase *et al.*, 1998; Vrijenhoek and Waypa, 2000). These processes allow efficient reduction of the As concentration at values lower than $10\,\mu g\,L^{-1}$ and do not produce toxic solid wastes. However, a high fouling potential of these membranes may hinder full scale applications.

Since As(V) can be adsorbed more strongly onto adsorbents than As(III), the oxidation of As(III) can be exploited and integrated with UF and MF systems at low pressure and adsorption/coagulation media for effective and low cost As removal.

2.2 MICROFILTRATION AND ULTRAFILTRATION

2.2.1 *General properties*

2.2.1.1 *Terminology*

Microfiltration (MF) and ultrafiltration (MF) are typical low-driven pressure membrane processes widely applied in various chemical and biochemical processes thanks to their advantages over traditional filtration methods. They are generally a thermal and simple in concept and operation and do not involve phase changes or chemical additives. Additionally, they are modular, easy to scale-up and characterized by low energy consumptions (Mulder, 1998).

In these processes fluids and solutes are selectively transported through a permselective barrier (membrane) under a hydrostatic pressure applied across it. As a result the feed solution is converted into two different streams: a solution containing all components which have permeated the membrane (permeate) and the remaining one containing all compounds rejected by the membrane (retentate) (Fig. 2.1). In most cases the feed flows tangentially to the membrane surface and the term 'cross-flow filtration' is used to describe such applications.

The separation mechanism in both MF and UF process is based on a sieving effect and particles are separated according to their dimensions although the separation is influenced by the interactions between the membrane itself and the particles being filtered. MF membranes generally have a symmetric structure and pores with diameter from 0.1 to $10\,\mu m$. Such membranes retain dispersed particles such as colloids, fat globules or cells: these particles are generally larger than those separated by UF and RO. Consequently, the osmotic pressure for MF is negligible and hydrostatic pressure differences used in MF are relatively small (in the range of 0.05–0.2 MPa).

The term UF is used when dissolved molecules or small particles with diameter not larger than $0.1\,\mu m$ are separated from a solvent or other low molecular weight compounds. Most UF membranes are asymmetric in structure with a dense active layer of 0.5–1 μm in thickness supported by a more porous support layer of greater thickness. Pore sizes in the skin layer are in the range 2–10 nm. Performance data of UF membranes are generally presented by membrane manufacturers in terms of molecular weight cut-off (MWCO), defined as the equivalent molecular weight of the smallest species that exhibit 90% rejection. The MWCO for UF membranes ranges between 2 and 300 kD (kilo Dalton).

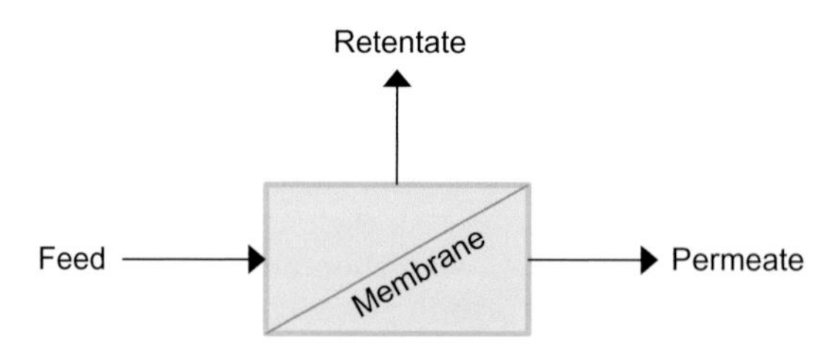

Figure 2.1. Schematic representation of a membrane process.

Typical rejected species include biomolecules, polymers and colloidal particles, as well as emulsions and micelles. Hydrostatic pressures are required to decrease with increasing MWCO and are generally between 0.1 and 0.5 MPa. In both MF and UF processes the filtration rate can be expressed by:

$$J = L_v \frac{\Delta p}{l} \tag{2.1}$$

where J is the permeate flux across the membrane, Δp the pressure difference between the feed and the permeate solution, l the membrane thickness and L_v is the hydrodynamic permeability of the membrane.

The separation characteristics of MF and UF membranes can be expressed in terms of membrane rejection or retention (R):

$$R = \left(1 - \frac{C_p}{C_f}\right) \cdot 100 \tag{2.2}$$

where C_p is the solute concentration in the permeate and C_f the solute concentration in the feed. Rejection values are between 0% (for solutes having the highest probability to pass through the membrane) and 100% (when solutes are completely retained by the membrane).

The volume reduction factor (VRF) in UF and MF processes is defined as the ratio between the initial feed volume and the volume of the resulting retentate given by:

$$VRF = \frac{V_f}{V_r} \tag{2.3}$$

where V_f and V_r are the volume of feed and retentate, respectively.

2.2.1.2 *Mode of operation*

MF and UF processes can be operated either in dead-end or in cross-flow configurations. Dead-end UF (Fig. 2.2a) is used on small-scale and laboratory applications: the feed flow is forced perpendicularly through the membrane causing a build-up of retained particles at the membrane surface and the formation of a cake layer. The thickness of the cake layer increases with the filtration time; therefore, the permeation rate decreases by increasing the cake layer thickness. The cross-flow mode (Fig. 2.2b) is largely used on medium- and large-scale processes. In this approach the fluid to be filtered flows tangentially to the membrane surface and permeates through the membrane due to the imposed transmembrane pressure (TMP) difference. This configuration allows minimizing the accumulation of solute and particles near the membrane surface. In addition, the recirculation of the retentate stream to the feed tank is facilitated and can be applied to mixing with fresh feed.

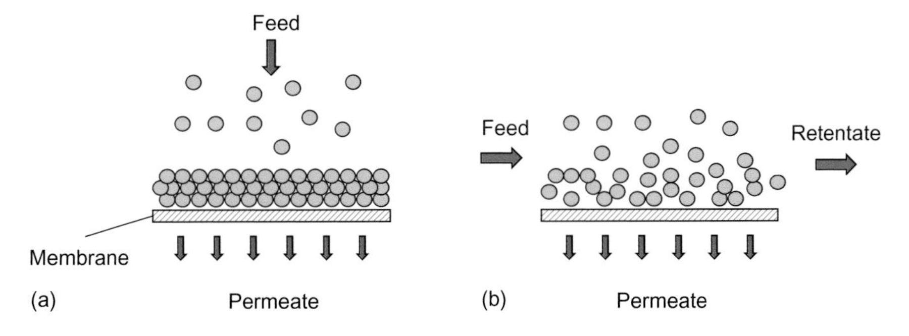

Figure 2.2. Mechanism of dead-end (a) and cross-flow (b) filtration.

Cross-flow MF and UF can be operated in different configurations, depending on the requirements of the process. Commonly used modes of operation are total recycle, batch concentration, feed-and-bleed and diafiltration (Cheryan, 1998; Ho and Sirkar, 1992).

In the total recycle configuration both permeate and retentate streams are recycled in the feed tank. This configuration is mainly used in research to measure the permeate flux at different operating conditions. In the batch mode, the retentate is recycled back to the feed tank while the permeate is collected separately. As a result, the concentration of particles increases with time. This configuration is usually restricted to small-scale operations. The feed-and-bleed configuration is commonly used at industrial level to obtain high concentration factors. The permeate is collected separately and the retentate is removed from the system when its final concentration is reached. Most of the retentate is recycled to maintain high tangential velocities through the membrane.

In the diafiltration operation the retentate is recycled in the feed tank and the fresh solvent (generally water) is added to the feed tank simultaneously with filtration. This configuration is typically used to improve the removal of dissolved solute species through the membrane system.

2.2.1.3 *Membranes and membrane modules*

MF and UF membranes can be polymeric or inorganic. Membrane materials must be chemical resistant to both feed and cleaning solutions, mechanically and thermally stable, and characterized by high selectivity and permeability. Polysulfone (PS), polyethersulfone (PES), polyamide (PA), cellulose acetate (CA), polyacrylonitrile (PAN), polytetrafluoroethylene (PTFE), poly(vinylidene F^-) (PVDF) and polypropylene (PP) are typical materials commonly used to cast the membrane. Alumina, zirconia and ceramic materials are usually used as inorganic materials.

Polymeric membranes, even if largely used in different industrial sectors, can operate in limited conditions of pH and temperature. Ceramic membranes offer a greater chemical, mechanical and thermal stability; on the other hand, the available pore size range is more limited.

The separation process in UF and MF systems is realized in proper devices known as membrane modules. A membrane module must be able to support the membrane, to minimize the concentration polarization phenomenon and to provide a large surface area in a compact volume.

The most common configurations of cross-flow modules are the plate-and-frame, spiral-wound, tubular and hollow-fiber types (Fig. 2.3).

Flat-sheet membranes are normally assembled in plate-and-frame devices together with porous support plates and spacers forming the feed flow channels. The feed solution is pressurized in the housing and forced across the membrane (Fig. 2.3a). The support plate provides a flow channel for the permeate that is collected from a tube on the side of the plate. Feed channel heights vary from 0.3 to 0.75 mm depending on the viscosity of the feed solution to be filtered.

In the spiral-wound configuration membranes are sandwiched together with feed flow channel spacers and the porous membrane support around a central permeate collecting tube (Fig. 2.3b). Commercial systems are about 1 meter long with diameters between 10 and 60 cm. Membrane areas can be in the range of 3–60 m^2. Spiral-wound membranes offer a good membrane surface/volume and low capital/operating cost ratios. Nevertheless, they cannot be mechanically cleaned and a feed pretreatment is required.

In tubular membrane modules membranes are cast on a relatively thick and mechanically strong porous support material. Tube diameters are typically in the range of 10–25 mm. The feed solution is fed through the tubular bundle while the permeate is collected on the shell side of the module (Fig. 2.3c). These systems allow an efficient control of the concentration polarization and membrane fouling phenomena and are easy to clean. However, as the tube diameter increases, they occupy a larger space and require high pumping costs.

Hollow fiber membrane modules (Fig. 2.3d), consisting of fibers with diameters of 0.001–1.2 mm, offer the highest packing density of all modules available on the market and can withstand relatively high pressures. However, the control of concentration polarization and membrane fouling is difficult and an extensive pretreatment of the feed solution is required in order to remove particles, macromolecules or other materials which can precipitate at the membrane surface.

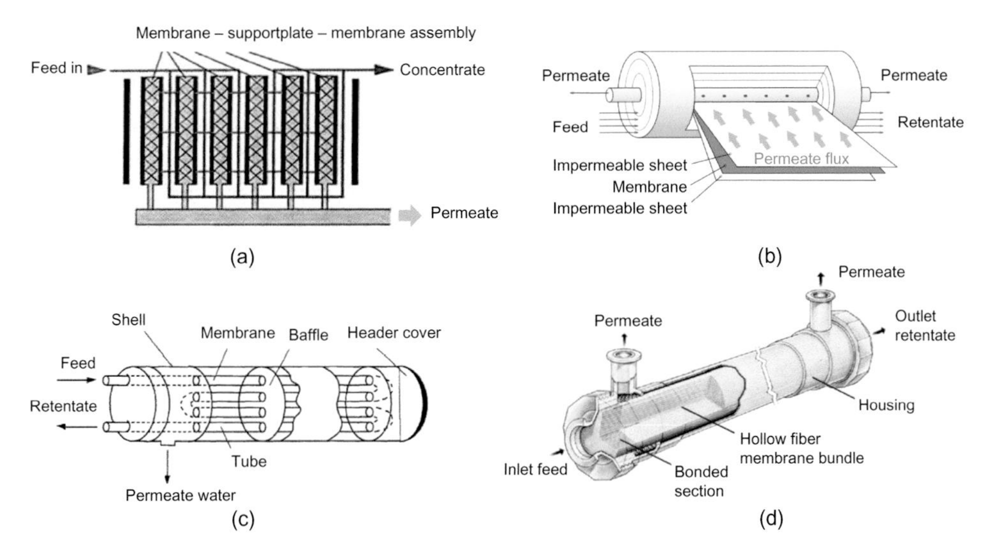

Figure 2.3. Schematic representation of plate-and-frame (a), spiral-wound (b), tubular (c) and follow-fiber (d) membrane modules.

2.2.1.4 *Concentration polarization and membrane fouling*

In both MF and UF processes the accumulation of rejected particles over the membrane surface under a given pressure leads to a development of a thin boundary layer adjacent to the membrane surface. This phenomenon, depicted in Figure 2.4, is known as concentration polarization (Fane, 1984; Rautenbach and Albrecht, 1986). The increased concentration of rejected solutes at the membrane surface leads to a reduction of the permeate flux, a modification of the rejection characteristics of the membrane, a decline in the driving force (as the osmotic pressure at the membrane-solution interface increases with concentration) and, often, the formation of a gel type layer over the membrane surface due to the precipitation of rejected solutes on the membrane.

The flux decay due to concentration polarization has been attributed to the hydrodynamic resistance to the solvent flow, in addition to the membrane resistance, generated by the accumulated solute on the membrane surface. Generally, small particles tend to form a dense particle cake layer, while large macromolecules provide a "gel" layer. According to another view the accumulation of solutes on the membrane surface results in a higher osmotic pressure with a decrease in the driving force (Cheryan, 1998).

The film theory assumes that the concentration gradient generated by the convective transport of solute to the membrane surface causes a diffusive transport of solute back into the bulk of the solution. At steady state the convective transport is counterbalanced by the diffusive flux. This is described mathematically by:

$$-D\frac{dC}{dx} = JC_{\mathrm{B}} \tag{2.4}$$

where D is the diffusion coefficient, dC/dx the concentration gradient over a differential element in the boundary layer, C_{B} the bulk concentration of rejected solute and J the permeate flux.

If D is assumed constant, this expression may be integrated to the boundary conditions to give:

$$J = \frac{D}{\delta}\ln\frac{C_{\mathrm{G}}}{C_{\mathrm{B}}} = k\ln\frac{C_{\mathrm{G}}}{C_{\mathrm{B}}} \tag{2.5}$$

where C_{G} is the gel concentration, δ the thickness of the boundary layer and $k = D/\delta$ is the mass transfer coefficient.

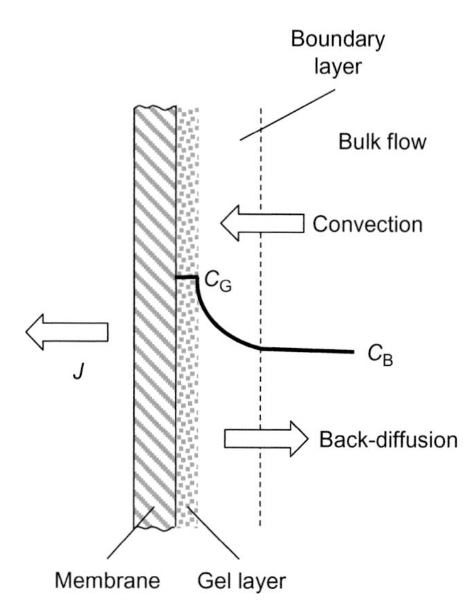

Figure 2.4. Schematic of concentration polarization (C_G = gel concentration; C_B = bulk concentration; J = permeate flux).

According to this model the permeate flux at steady-state is independent of the imposed pressure drop and it is controlled by the concentration polarization boundary layer. The increasing pressure drop results in a thicker solute layer until convection and diffusion in the boundary layer will balance again.

A more simple approach to describe the membrane flux in case of a gel or a cake layer formation is based on the use of the "resistance-in-series" concept. According to this model, the osmotic pressure of the feed solution can be neglected and the permeate flux can be expressed as:

$$J = \frac{\Delta p}{\mu(R_m + R_f + R_p)} \tag{2.6}$$

where μ is the viscosity of the solvent, Δp the transmembrane pressure, R_m the membrane resistance, R_f the fouling layer resistance and R_p the polarization layer resistance (Fane, 1983).

Concentration polarization phenomena in MF and UF processes can be minimized by using different approaches including increasing the cross flow velocity, insertion of turbulence promoters, chemical modification of the membrane surface, application of electrical fields influencing the charge on macromolecules, use of ultrasonics inducing cavitation at the surface and inert gas inducing turbulence.

The term *membrane fouling* is used to describe a long term flux decline caused by the interactions between retained particles and the membrane surface and/or membrane pores. It may occur due to a concentration polarization layer development over the membrane surface, the formation of a cake layer and/or a blockage of the membrane pores.

Solute properties (conformation, hydrophobic interactions, charge, etc.), operating conditions (cross-flow velocity, pressure and temperature) and membrane material are all factors which affect membrane fouling.

The consequences of membrane fouling are in terms of higher capital costs due to the lower average permeate flux, reduction of operating life of the membrane due to the use of cleaning agents to restore membrane flux, change of the effective sieving and transport properties of the membranes (Cheryan, 1998).

Methods and strategies to reduce membrane fouling include feed pretreatment systems, selection of appropriate membranes and membrane modules, change of membrane properties, selection of proper operating conditions and membrane cleaning.

Although the flux decline is a typical aspect involved in concentration polarization and membrane fouling, there are some substantial differences between these phenomena. Concentration polarization is a reversible process which takes place over a few seconds. It can be easily controlled by decreasing the TMP, lowering the feed concentration or increasing the cross-flow velocity. In membrane fouling, the flux decline is irreversible and takes place over many minutes, hours or days. It is more difficult to describe and to control experimentally. In addition, a continuous flux decline can often be observed.

2.3 ARSENIC REMOVAL BY USING MICROFILTRATION

MF membranes can remove only particulate forms of As in water since their pore size is too large to remove dissolved or colloidal species of As (Amy *et al.*, 2000). Unfortunately, the content of As as a particulate in water is very low. Coagulation and flocculation processes prior to MF can increase the particle size of As bearing species and, consequently, improve the As removal efficiency.

2.3.1 *Combined process coagulation/MF*

A combined coagulation/MF system to remove As from groundwater in Albuquerque (New Mexico) was investigated by Ghurye *et al.* (2004). In this approach an iron-based coagulant, such as ferric chloride, is added in water to form a $Fe(OH)_3$ precipitate with a net positive charge on the surface. At pH values between 4 and 10 arsenite is neutral in charge, while arsenate is negatively charged: therefore it can be adsorbed onto the positively charged precipitate by surface complexation. A MF membrane with a nominal pore size of 0.2 μm is used in the following step to remove the adsorbed As. The process allows production of treated water containing less than 2 μg As L^{-1} (starting from a feedwater with 40 μg As L^{-1}) using either 7 mg L^{-1} of ferric chloride, without a pH reduction, or a smaller dose of 1.9 mg L^{-1} Fe after the addition of sulfuric acid to reduce the pH value at 6.4. No increase in TMP was observed on a pilot-scale process after five days of continuous operation as a consequence of a low fouling index of the membrane.

When compared with ion exchange and activated alumina adsorption, the combined coagulation/MF system presents lower capital and maintenance costs. The principal differences are related to the large salt requirement for the ion exchange process and to the necessity of reducing the pH to 6 for activated alumina adsorption, followed by base addition to stabilize water (Chwirka *et al.*, 2000).

The combined coagulation/NF has been recognized by the USEPA as an emerging technology for the removal of As from drinking water. A summary of the process concepts, chemistry and design considerations for the use of this technology is reviewed by Chwirka *et al.* (2004).

The removal efficiency for arsenite is poor if compared to that for arsenate since arsenite exists as a neutral species and the coagulation processes rely upon ionic interactions. Consequently, a complete oxidation of arsenite to arsenate is needed.

According to the results of Brandhuber and Amy (1998) the most important variables controlling As removal are pH and concentration of ferric chloride. In particular, the As removal increases by increasing the coagulant dose, while the percentage of As removal decreases by increasing the pH value in the range from 6 to 9. The apparent size of the As bearing floc remains in the submicron range for coagulant doses (as mg L^{-1} $FeCl_3$) between 5 and 20 mg L^{-1}. For coagulant doses of 25 mg L^{-1} most As is removed as flocs of apparent size greater than 1 μm. For a MF membrane with a pore size of 0.2 μm the following equation was found to describe empirically the hyperbolic relationship between dose and As removal:

$$\%\ \text{As removal} = 100 \cdot \frac{k \cdot \text{Dose}}{1 + k \cdot \text{Dose}} \tag{2.7}$$

where $k = 0.332$ L mg^{-1} and Dose is the $FeCl_3$ dose in mg L^{-1}.

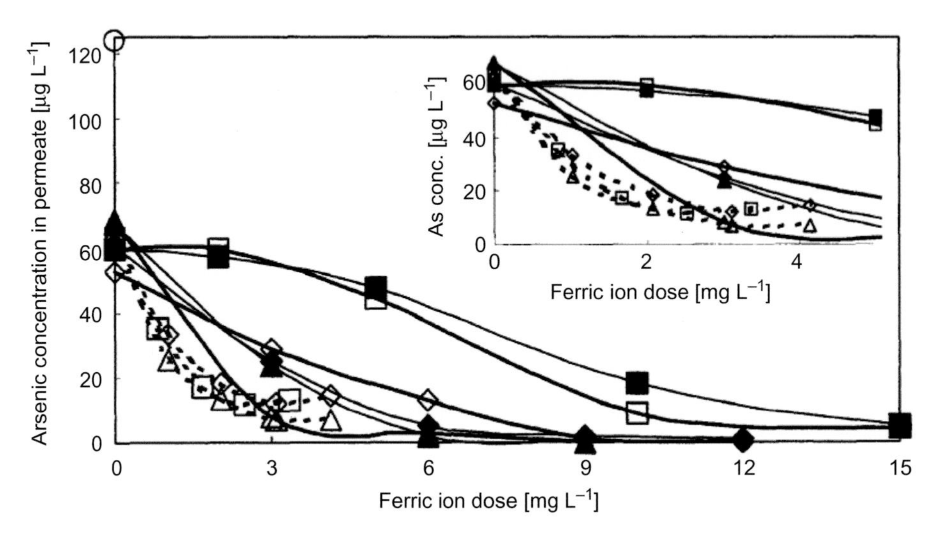

Figure 2.5. Effect of raw water pH on the variation of residual arsenic concentration with ferric ion dose. After coagulation, the suspension was vacuum filtered using 0.22 μm pore size membrane disk. ■ + thin solid line: US water, pH 8.7, $Fe_2(SO_4)_3$ as coagulant; ♦ + thin solid line: US water, pH 6.8, $Fe_2(SO_4)_3$ as coagulant; ▲ + thin solid line: US water, pH 6.2, $Fe_2(SO_4)_3$ as coagulant; □ + thick solid line: US water, pH 8.7, $FeCl_3$ as coagulant; ◇ + thick solid line: US water, pH 6.8, $FeCl_3$ as coagulant; △ + thick solid line: US water, pH 6.2, $FeCl_3$ as coagulant; O unfiltered Bangladesh water, pH 7.5; ◇ + dashed line: Bangladesh water, pH 6.8, $Fe_2(SO_4)_3$ as coagulant; △ + dashed line: Bangladesh water, pH 6.2, $Fe_2(SO_4)_3$ as coagulant; □ + dashed line: Bangladesh water, pH 6.8, $FeCl_3$ as coagulant (Wickramasinghe *et al.*, 2004, with permission from Elsevier).

A similar relationship, but with a k value of 0.496, was found by Brandhuber and Amy (1998) in pilot tests performed by using a Memcor 4M1W 0.2 μm MF unit and coagulant doses of $FeCl_3$ from 2 to $10\,mg\,L^{-1}$. A dose of $7.0\,mg\,L^{-1}$ as $FeCl_3$, a permeate flux of about $170\,L\,m^{-2}\,h^{-1}$ and a 90% recovery were identified as optimum operating conditions for the pilot plant. In these conditions, an average As rejection and turbidity reduction of 84% and 64%, respectively, were obtained. Membrane fouling was successfully controlled by air backwashes of the membrane at 15 minute intervals.

Wickramasinghe *et al.* (2004) evaluated the removal of As from USA and Bangladesh groundwater by using a combined approach coagulation/MF. Ferric chloride or ferric sulfate was used as coagulant. In addition, a cationic polyelectrolyte (CY 2461, Cytec Industries, Stamford, CT) was also tested as a coagulant aid (doses were of 0.02 and $0.3\,mg\,L^{-1}$). MF was performed by using hollow-fiber membranes from A/G Technology (Needham, MA) with nominal pore size of 0.1 μm. Results of bench-scale experiments indicated that the As removal is highly dependent on the raw water quality and the used coagulants gave efficient results; however, the use of ferric sulfate led to a lower residual turbidity in the treated water. The addition of polyelectrolyte as a coagulant aid improved the permeate flux but had no effect on the As residual concentration. In a pH range of 6.2–8.7 the As removal was improved by decreasing the pH value (Fig. 2.5). This result can be explained assuming that when pH is lowered, the As adsorption is increased leading to an increased particle size (Jain and Loeppert, 2000; Meng *et al.*, 2000). Consequently, also the membrane rejection towards precipitate particles at a given ferric ion dose is increased by decreasing pH. Therefore, pH adjustment may be necessary in order to reduce the ferric ion dose required.

Similar results were obtained by Han *et al.* (2002) in a combined system flocculation/MF in which ferric chloride and ferric sulfate were used as flocculants and mixed esters of cellulose acetate and cellulose nitrate with pore sizes of 0.22 and 1.2 μm as MF membranes. Flocculation prior to MF led to significant As removal in the permeate with a consequent reduction of its turbidity.

Recently the combination of the electrocoagulation (EC) process and MF was found to be effective in removal 98.9% of As from a feed solution containing 200 μg As L^{-1} in presence of fluoride and iron contaminant (Ghosh *et al.*, 2011).

MF processes are characterized by lower energy requirement and higher fluxes when compared with NF and RO. Consequently, the combination of flocculation with MF represents a cost effective method to reduce the As content of drinking water.

2.4 ARSENIC REMOVAL BY USING ULTRAFILTRATION

UF membranes, similarly to MF membranes, are not able to remove dissolved or colloidal species of As in water due to their large pore size. However, significant removal efficiencies can be achieved by using UF membranes negatively charged, colloid-enhanced UF (CEUF) and electro-ultrafiltration (EUF).

2.4.1 *Negatively charged UF membranes*

UF membranes with electric repulsion show a better As removal efficiency when compared to UF membranes with only pore-size dependent sieving. The influence of membrane charge on the As removal efficiency was investigated in bench-scale tests by Amy *et al.* (1998). Negatively charged GM2540F UF membranes (supplied by GE Osmonics, Minnetonka, MN) gave a 63% rejection of As(V) at neutral pH but a very low rejection at acidic pH. A poor rejection for both As(III) and As(V) species was observed by using an uncharged UF membrane (FV2540). So the high removal rate of As(V) was attributed to the electrostatic interaction between As ions and the negatively charged membrane surface. Pilot-scale studies showed As removal of about 70% in groundwater with high dissolved organic compounds and lower rejections (about 30%) in those with low dissolved organic compounds. This behavior was attributed to the adsorption of natural organic matter which reduces the membrane surface charge increasing the repulsion towards negatively charged As.

The influence of operating conditions and water quality on the As rejection of negatively charged UF membranes was evaluated by Brandhuber and Amy (2001). Bench-scale experiments were carried out by using a flat-sheet cross flow cell equipped with a thin film composite sulfonated polysulfone membrane having a MWCO of 8 kDa (GE Osmonics, GM). The observed trends of As(V) rejection were in agreement with the Donnan exclusion mechanism. In particular, a reduction of As(V) rejection was observed by increasing the bulk As(V) concentration and the ionic strength of the feed solution. The presence of co-occurring divalent ions, such as Ca^{2+} and Mg^{2+} reduced also the As(V) rejection. This phenomenon was attributed to the formation of ion pairs between counter ions and the fixed charged group in the membrane matrix, which locally neutralizes the membrane charges. According to the Donnan equilibrium, the As(V) rejection slightly increased with increasing flux at constant recovery (Bhattacharyya and Grieves, 1978). An increasing recovery at constant permeate flux produced a substantial decrease in apparent As rejection due to a decrease in membrane cross-flow velocity. According to the concentration polarization phenomenon, a decrease in cross-flow velocity results in a thicker boundary layer, decreased solute back transport and a greater concentration of As at the membrane surface. A slight reduction of As rejection by increasing the temperature in the range 20–40°C was attributed to the increased diffusivity of As with temperature which in turn increases the diffusive transport of As across the membrane. The presence of natural organic matter (NOM) improved the As(V) rejection in presence of divalent cations. This phenomenon can be attributed to the complexation capacity of divalent ions, such as Ca^{2+} and Mg^{2+}, with organic matter (Mathuthu and Ephraim, 1993) with a consequent modification of free ion distribution at the membrane surface. This leads to a reduction of the equilibrium partitioning of As ions into the membrane reducing their transport through the membrane and, consequently, the As rejection. In addition, the adsorption of organic matter onto the membrane surface leads to a formation of a negatively charged layer in the adjacent membrane layer increasing the rejection of negatively charged As species.

The applicability of polyacrylonitrile (PAN)-based negatively charged UF membrane for As removal was demonstrated for the first time by Lohokare *et al.* (2008). The surface of flat-sheet PAN-based UF membranes was hydrolyzed by using NaOH in cross-flow mode: this approach led to the reduction in pore size (as demonstrated by the reduction in water flux and the increased rejection towards proteins and polyethylene glycol) due to the formation of carboxylate ($-COO^-$) groups on both membrane and pore wall surfaces and an increased membrane hydrophilicity. The MWCO after the NaOH treatment was found to be of about 6 kDa. These modified membranes showed excellent As(V) rejections (close to 100%) when simulated solutions containing 50 $\mu g\,L^{-1}$ of As in pure water were used as feed samples. For these samples the rejection coefficient was independent of cross-flow velocity and TMP. For feed samples containing 1000 $mg\,L^{-1}$ of As the rejection was between 40 and 65% depending on the cross-flow velocity and TMP and, consequently, on concentration polarization phenomena.

The effect of foulants, such as ovalbumin, humic acid and egg white, on the As(V) rejection of modified PAN UF membranes has been recently investigated (Agarwal *et al.*, 2013). Results indicate a decreasing in As(V) rejection with increasing concentration of proteins and humic acid. In addition the As(V) rejection was affected by the total quantity of proteins and not by the variety of proteins in the feed solution.

2.4.2 *Colloid-enhanced ultrafiltration (CEUF)*

Colloid-enhanced ultrafiltration (CEUF) is a separation technique based on the use of colloids able to bind multivalent metal ions by electrostatic interactions. The colloidal solution is then filtered under pressure through a UF membrane with a pore size smaller than the size of the colloid, producing a purified water stream (permeate) and a concentrated stream containing almost all of the colloid and metal ions (retentate) (Dunn *et al.*, 1989). CEUF can be distinguished in micellar-enhanced ultrafiltration (MEUF), if the colloidal species is a micelle-forming surfactant, and in polyelectrolyte-enhanced ultrafiltration (PEUF) when the colloidal species is a polyelecrolyte.

2.4.2.1 *Micellar-enhanced ultrafiltration (MEUF)*

Micellar-enhanced UF (MEUF) is a technology that employs surfactant micelles to solubilize inorganic and organic pollutants from effluent streams (De and Mondal, 2012). It is particularly effective for removal of metal ions (Juang *et al.*, 2003; Liu and Li, 2005; Rahmanian *et al.*, 2010), small amounts of organic substances (Adamczak *et al.*, 1999; Dunn *et al.*, 1985; Sabatè *et al.*, 2002) and anionic pollutants such as chromate, nitrate and phosphate (Baek and Yang, 2004; Morel *et al.*, 1997).

Surfactants are usually organic compounds containing both hydrophobic groups acting as their *tails* and hydrophilic groups acting as their *heads*. Typically, the tail consists of a hydrocarbon chain which can be branch, linear or aromatic. In the bulk aqueous phase, surfactants form micelles where the hydrophobic tails form the core of the aggregate and the hydrophilic heads are in contact with the surrounding liquid (Fig. 2.6).

Other types of aggregates such as spherical or cylindrical micelles or bilayers can be formed depending on the balance of the sizes of the hydrophobic tail and the hydrophilic head.

Surfactants are generally classified according to the polar head group as: non-ionics (if they have no charge groups in their head), anionics (if the head carries a negative net charge), cationics (if the head carries a positive net charge) and zwitterionics (if the head contains two oppositely charged groups).

When a cationic surfactant is added to contaminated water above its critical micelle concentration (CMC), it forms micelles positively charged on the surface which can adsorb anionic As

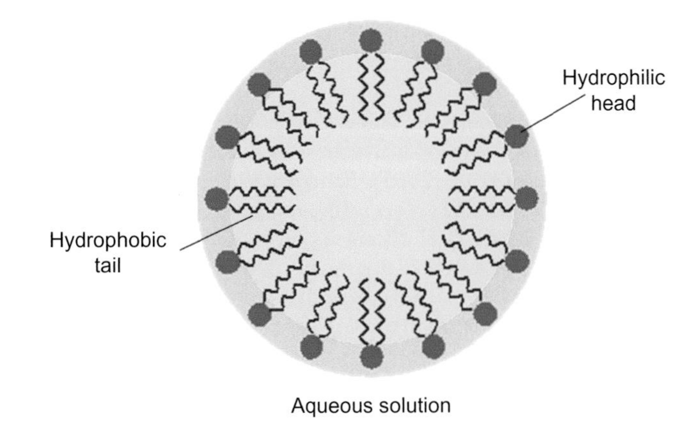

Figure 2.6. Schematic representation of spherical micelles.

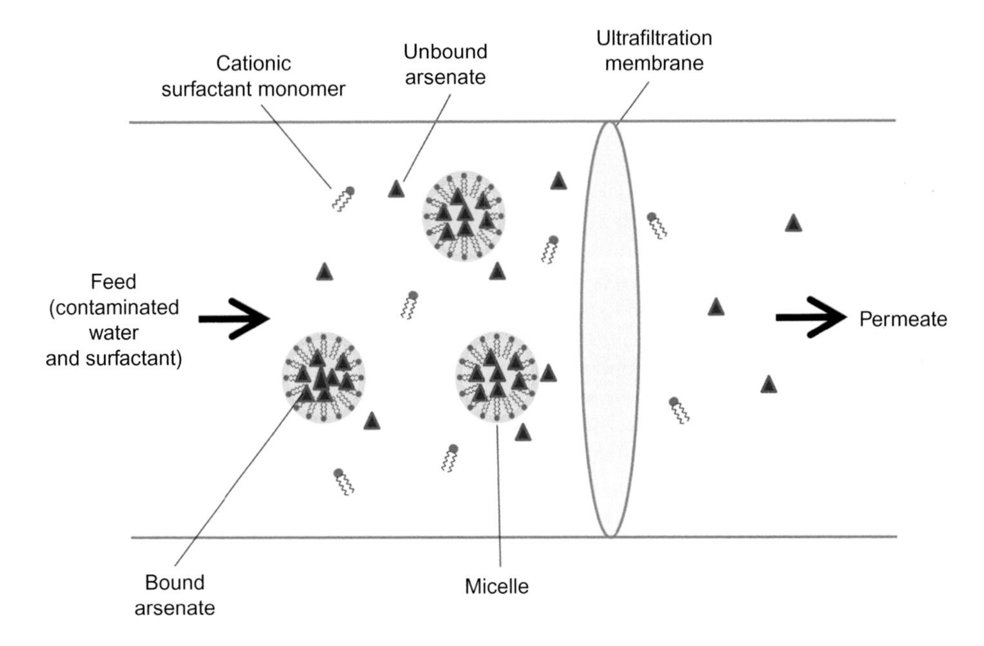

Figure 2.7. Schematic of As removal process by MEUF.

species (arsenite or arsenate) by electrostatic interaction. These micelles can be efficiently separated from aqueous streams by using UF membranes whose pores are too small to allow their permeation (Fig. 2.7).

Gecol *et al.* (2004) investigated the removal of As(V) from water by using flat sheet UF membranes and cetylpyridinium chloride (CPC) as cationic surfactant. In particular, regenerated cellulose (RC) membranes with a MWCO of 10 kDa and polyethersulfone (PES) membranes with MWCO of 5 and 10 kDa were used. When membranes were used without surfactant micelles PES membranes exhibited a lower As removal than RC membranes. This was attributed to the negatively charged surface of RC membranes and, consequently, to a Donnan exclusion mechanism. However, in both cases the As concentration in the permeate stream was higher than the MCL of 10 $\mu g\,L^{-1}$. The addition of CPC (10 mM) reduced the As concentration in the permeate of all tested membranes well below the MCL.

Both As removal and permeate fluxes were influenced by membrane materials, MWCO and pH. The maximum As removal was obtained with 5 kDa PES membranes at pH 5.5 and 10 kDa RC membranes at pH 8.

The presence of co-occurring inorganic solutes, such as HCO_3^-, HPO_4^{2-}, H_4SiO_4 and SO_4^{2-} species, did not affect the As(V) removal efficiency when a PES 5 kDa membrane was used in combination with CPC (Ergican *et al.*, 2005). Permeate fluxes decreased by increasing the As concentration of the feed water and the co-occurring inorganic solute concentration.

CPC exhibited the highest As removal efficiency (96%) when used in combination with RC membranes of 3 and 10 kDa (YM3 and YM10, Amicon, USA). The removal of arsenate with hexadecyltrimethylammonium bromide (CTAB), benzalkonium chloride (BC) and octadecylamine acetate (ODA) at a surfactant concentration of 10 mM was of 94, 81 and 55%, respectively (Iqbal *et al.*, 2007).

2.4.2.2 *Polyelectrolyte-enhanced ultrafiltration (PEUF)*

Water soluble polymers with ion exchange properties can be efficiently used to remove ions from aqueous medium. For example, polymeric electrolytes containing quaternary ammonium groups, as ion-exchanged cationic groups, have been extensively investigated for their metal ion binding properties (Rivas *et al.*, 2006a; 2006b). The arsenate retention with two cationic water-soluble polymers containing tetra alkylammonium groups (poly[2-(acryloyloxy)ethyl]trimethylammonium chloride and poly[2-(acryloyloxy)ethyl]trimethylammonium methyl sulphate) was investigated by Rivas *et al.* (2007) by using the liquid-phase polymer-based retention (LPR) technique. Results demonstrated a greater retention property for the cationic soluble polymer containing chloride counteranions at pH 8.

The addition of water-soluble polymers followed by ultrafiltration, named as polyelectrolyte-enhanced ultrafiltration (PEUF), can be efficiently exploited to remove ionic species from aqueous solutions. This process is based on the use of a polyelectrolyte having an opposite charge to that of the target ions and the formation of macromolecular complexes between pollutant ions and polymer due to electrostatic attractions. These complexes are too large to pass through a UF membrane so they are retained in the retentate streams. Examples of separation of both cationic and anionic metal ions by PEUF have been extensively reported (Christian *et al.*, 1995; Tabatabai *et al.*, 1995a; Tangvijitsri *et al.*, 2002).

The use of PEUF for the removal of As anions from water has been investigated by Pookrod *et al.* (2004). In this approach, a cationic polyelectrolyte, poly(diallyldimethyl ammonium chloride) (poly-DADMAC), was used to bind anionic As species in order to form macrocomplexes which can be retained by a UF membrane (10 kDa RC acetate membrane, Millipore, Bedford, MA). The repeating unit of the polymer is $(H_2CCHCH_2)_2N(CH_3)_2Cl$. A schematic diagram of the process is depicted in Figure 2.8.

An increasing of the [poly-DADMAC]/[As] ratio produced an increasing of the number of positively charged sites on the polymer per unit volume and, consequently, of the fraction of bound As anions. Therefore, the As rejection is enhanced by increasing the polymer concentration.

The As rejection decreased by increasing the feed salt concentration and the valence of the added anion. This phenomenon can be attributed to the competition between arsenate and other anions (such as phosphate, silicalite, carbonate commonly present in water) for binding sites on the polymer. A similar behavior has been also observed in the As removal by using ion-exchange resins containing ammonium groups (Berdal *et al.*, 2000).

The As rejection increased by increasing the pH. In particular, an increase in pH from 6.5 to 8.5 produced an increase in As rejection from 99.06 to 99.95% due to an increasing of the ratio $HAsO_4^{2-}/H_2AsO^{4-}$ in the feed solution so improving the As binding capacity to the polymer.

The polymer concentration at which the permeate flux is zero, defined as gel point concentration, was found to be 655–665 mM (approximately 5.98–6.07 wt%). These values allow high water recoveries (higher than 99%) to be obtained. An advantage of the PEUF system in the

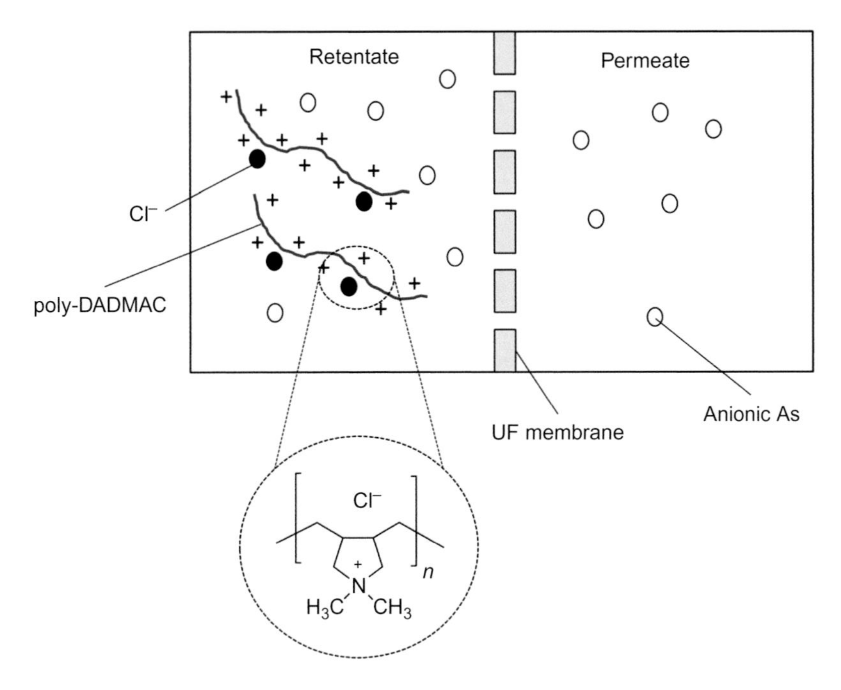

Figure 2.8. Schematic diagram of PEUF to remove anionic arsenic from water.

removal of As, if compared with other pollutants, is that the As feed concentration is often very low (lower than $100\,\mu g\,L^{-1}$).

Pookrod *et al.* (2004) estimated that for a [poly-DADMAC]/[As] ratio of 100 and an As feed concentration of $100\,\mu g\,L^{-1}$ the retentate could be concentrated by a factor of 547 (up to 72.9 mM), when the relative flux is reduced to 0.4, with a permeate/feed volume of 0.998.

PEUF shows a greater potential when compared with a combination of ion exchange and UF for the removal of hardness ions as well as viruses, bacteria and pyrogen (Tabatabai *et al.*, 1995b). The formation of polarized layers due to the presence of polyelectrolites is one of the disadvantages of the PEUF process since it has a detrimental effect on permeate fluxes and total costs. The combination of colloids as well as the development of improved turbulence promoters has been suggested as a source of further sustainable growth of the PEUF process.

2.4.3 *Electro-ultrafiltration (EUF)*

As previously reported, the pore size of MF and UF membranes is too small to remove arsenite and arsenate species whose molecular weights are 126 and $142\,g\,mol^{-1}$, respectively. Typical As(III) and As(V) rejections of UF membranes (Brandhuber and Amy, 1998) are 5 and 40%, respectively. NF and RO processes, based on the use of dense membranes, allow obtaining higher separation efficiencies. The different charge characteristics of As(III) and As(V) species are very important factors that should be carefully considered when these membrane technologies are employed for As removal. At neutral pH the predominant species for As(V) are $H_2AsO_4^-$ and $HAsO_4^{2-}$ which means that As(V) exists as an anion at a typical pH in natural water (pH 5–8), whereas in this range of pH As(III) is mainly present as uncharged species (H_3AsO_3) and, therefore, is less efficiently rejected by RO or NF membranes (Amy *et al.*, 1998; Brandhuber and Amy, 2001; Geucke *et al.*, 2009; Uddin *et al.*, 2007a).

The role of Donnan exclusion in the As removal from water by using loose NF membranes was investigated by Seidel *et al.* (2001): they found that the removal of As(V) increased from

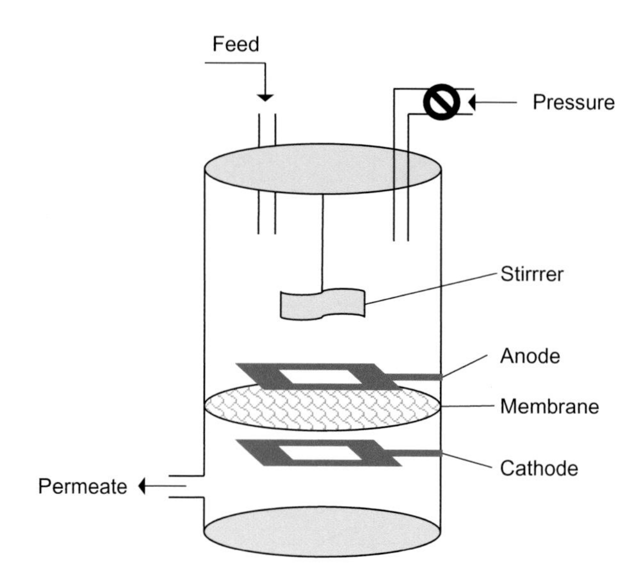

Figure 2.9. Schematic diagram of electro-ultrafiltration (adapted from Saxena *et al.*, 2007).

60 to 90% as the As feed water concentration was increased from 10 to 316 $\mu g\,L^{-1}$; in a similar range of feed concentration the rejection of As(III) decreased from 28 to 5%.

According to these results, the performance of UF membranes in the removal of As from water can be enhanced through the application of an electric field across the membrane in order to attract the As charged species. In this process, called electro-ultrafiltration (EUF), the electric field acts as an additional driving force to the TMP (Charcosset, 2012; Huotari *et al.*, 1999; Saxena *et al.*, 2009). The electric gradient is generally applied by two parallel electrodes positioned on either side to the UF membrane. In most cases platinum electrodes are inserted into the feed and permeate channels (Saxena and Shahi, 2007). When an electric field is applied to the system a displacement of charged species towards the electrode with the opposite sign occurs (Fig. 2.9).

EUF has been largely investigated for separation or concentration of protein solutions (Sung *et al.*, 2007; Zhou *et al.*, 1995). In addition, it is an effective method to decrease the gel layer formation on the membrane surface and to increase the filtration flux due to electrokinetic phenomena such electrophoresis and electroosmosis (Weber and Stahl, 2003).

The removal of As and humic substances (HSs) from water by using EUF was investigated by Weng *et al.* (2005). The laboratory scale EUF system was equipped with a polyacrylonitrile (PAN) UF membrane with a MWCO of 100 kDa (GE Osmonics). Results indicated that the As(V) rejection in presence of humic acid increased from 30% to more than 90% when an electric voltage was applied. On the other hand, the As(III) rejection was lower than 20%, independently on the presence of humic substances, even with the addition of electrical voltages (Fig. 2.10).

Since As(III) is non-ionic at a neutral pH it is not influenced by the electrical field. However, the removal of As(III) can be enhanced through the oxidation of As(III) to As(V) (Bissen and Frimmel, 2003) or by increasing the pH of water thus leaving As(III) negatively charged (Kang *et al.*, 2000).

The EUF system was also tested on two groundwater samples coming from the I-Ian County in the northeastern part of Taiwan (Chaung-Hsieh *et al.*, 2008). The removal of As in absence of an electrical voltage was in the range of 1–14%. The application of an electrical voltage of 25 V to the UF system reduced the total As concentration in both groundwater samples by over 79%. The association between As(III) species and dissolved organic matter was suggested as a possible factor enhancing the As removal.

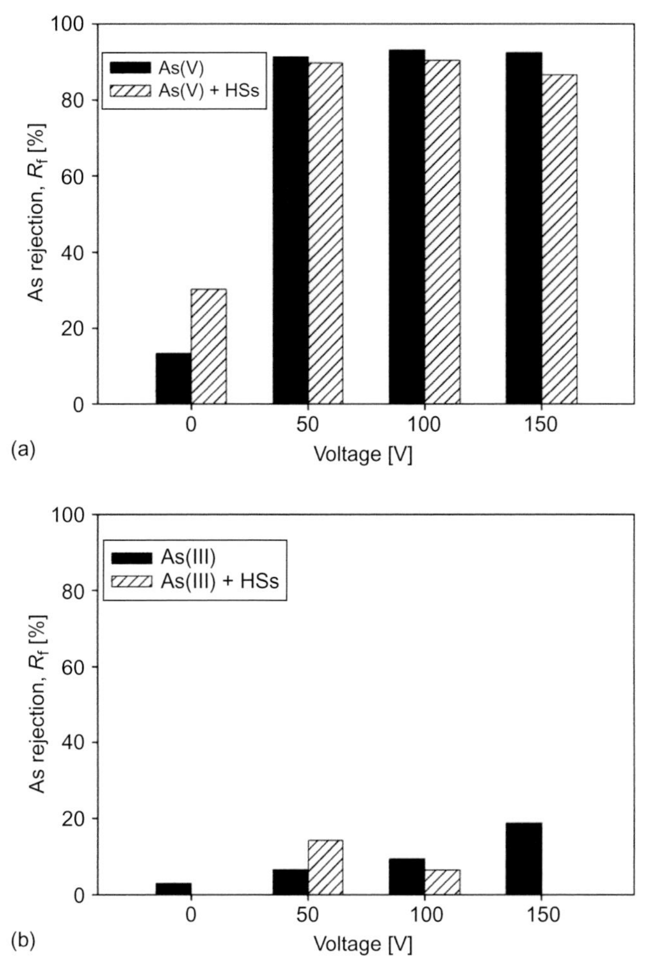

Figure 2.10. Removal of synthetic (a) As(V) and (b) As(III) from water by EUF at different voltages (pH = 6) (Weng *et al.*, 2005, with permission from Elsevier).

2.5 CONCLUSIONS

Low pressure membrane technologies, such as MF and UF, have been demonstrated to be effective in the removal of As from water when combined with coagulation/flocculation methodologies or colloidal species such as micelle-forming surfactants and polyelectrolytes.

The removal of As by MEUF depends on the As concentration and surfactants, solution pH, ionic strength, and parameters related to membrane operation. The surfactant may account for a large portion of operating costs: therefore, the recovery and reuse of the surfactant is a key factor to make the process economically feasible and to avoid secondary pollution. However, the combination of high MWCO membranes and low surfactant concentrations can benefit the overall process economics for the lower membrane area requirement (due to greater flux) and the reduced surfactant consumption.

The use of negatively charged membranes and the application of an electric field acting as an additional driving force to the transmembrane pressure across UF membranes have been also successfully explored for the removal of As from water sources. In these approaches the removal efficiency of As(V) is higher if compared with the removal of As(III). Therefore, the use of oxidizing agents (i.e., $KMnO_4$, NaClO and ozone) improves the As removal in raw waters where As(III) is predominant. Since oxidizing agents can damage MF and UF membranes, research efforts can be addressed to explore the integration of microorganisms able to transform arsenate to arsenite (biooxidation) with MF or UF membranes.

Membrane properties and operating conditions (pH, temperature, pressure) affect the As removal efficiency and the operating costs. Therefore, they should be carefully selected and optimized for each water source.

Complexing agents have been proven to achieve selective separation of As with low energy requirements. The main parameters affecting PEUF are the polymer type, the ratio of As to polymer, pH and existence of other metal ions in the solution. Advantages of PEUF include high removal efficiency and high binding selectivity.

The combination of colloids as well as the development of improved turbulence promoters appears as an attractive way to improve the performance of the PEUF process.

NOMENCLATURE

C	concentration [mol m^{-3}, mol L^{-1}]
D	diffusion coefficient [m^2 s^{-1}]
J	permeate flux [m s^{-1}]
k	mass transfer coefficient [m s^{-1}]
l	membrane thickness [m]
L_v	hydrodynamic permeability [m^2 Pa^{-1} s^{-1}]
p	pressure [Pa]
R	rejection [–]
R_m	membrane resistance [m^{-1}]
R_f	fouling resistance [m^{-1}]
R_p	polarization layer resistance [m^{-1}]
V	volume [m^3]
VRF	volume reduction factor [–]
x	directional co-ordinate [m]

Greek letters

δ	boundary layer thickness [m]
Δ	difference [–]
μ	viscosity [Pa s]

Subscripts

B	bulk
f	feed
G	membrane wall (gel)
p	permeate
r	retentate

Abbreviations

As	arsenic
CA	cellulose acetate
MF	microfiltration
MWCO	molecular weight cut-off
NF	nanofiltration

PA	polyamide
PAN	polyacrylonitrile
PES	polyethersulfone
PP	polypropylene
PS	polysulfone
PTFE	polytetrafluoroethylene
PVDF	poly(vinylidene fluoride)
RO	reverse osmosis
TMP	transmembrane pressure
UF	ultrafiltration
USEPA	US Environmental Protection Agency

REFERENCES

Adamczak, H., Materna, K., Urbanski, R. & Szymanowski, J. (1999) Ultrafiltration of micellar solutions. *Journal of Colloid and Interface Science*, 218 (2), 359–368.

Agarwal, G.P., Karan, R., Bharti, S., Kumar, H., Jhunjhunwala, S., Sreekrishnan, T.R. & Kharul, U. (2013) Effect of foulants on arsenic rejection via polyacrylonitrile ultrafiltration (UF) membrane. *Desalination*, 309, 243–246.

Amy, G.L., Edwards, M., Benjamin, M., Carlson, K., Chwirka, J., Brandhuber, P., McNeill, L.S. & Vagliasindi, F. (1998) Arsenic treatability options and evaluation of residuals management issues. American Water Works Association Research Foundation (AWWARF) report, April 1998, Denver, CO.

Amy, G.L., Edwards, M., Benjamin, M., Carlson, K., Chwirka, J., Brandhuber, P., McNeill, L.S. & Vagliasindi, F. (2000) Arsenic treatability options and evaluation of residuals management issues. American Water Works Association Research Foundation (AWWARF) report No. 90771, Denver, CO.

Baek, K. & Yang, J.W. (2004) Cross-flow micellar-enhanced ultrafiltration for removal of nitrate and chromate: competitive binding. *Journal of Hazardous Materials*, 108 (1–2), 119–123.

Berdal, A., Verrie, D. & Zaganiaris, E. (2000) Removal of arsenic from potable water by ion exchange resins. In: Greig, J.A. (ed.) Ion exchange at the millenium. *Proceedings of IEX 2000*. Imperial College Press, London, UK. pp. 142–149.

Bhattacharyya, D. & Grieves, R. (1978) Charged membrane ultrafiltration. In: Li, N. (ed.) *Recent developments in separation science*. CRC Press, Boca Raton, FL.

Bissen, M. & Frimmel, F. (2003) Arsenic – a review. Part II: Oxidation of arsenic and its removal in water treatment. *Acta Hydrochimica et Hydrobiologica*, 31 (2), 97–107.

Bodzek, M., Konieczny, K. & Kwiecinska, A. (2011) Application of membrane processes in drinking water treatment – state of art. *Desalination and Water Treatment*, 35 (1–3), 164–184.

Brandhuber, P. & Amy, G. (1998) Alternative methods for membrane filtration of arsenic from drinking water. *Desalination*, 117 (1), 1–10.

Brandhuber, P. & Amy, G. (2001) Arsenic removal by charged ultrafiltration membrane – influences of membrane operating conditions and water quality on arsenic rejection. *Desalination*, 140 (1), 1–14.

Cakmakci, M., Baspinar, A.B., Balaban, U., Uyak, V., Koyuncu, I. & Kinaci, C. (2009) Comparison of nanofiltration and adsorption techniques to remove arsenic from drinking water. *Desalination and Water Treatment*, 9 (1–3), 149–154.

Charcosset, C. (2012) *Membrane processes in biotechnology and pharmaceutics*. Elsevier, Oxford, UK.

Chaung-Hsieh, L.H., Weng, Y.H., Huang, C.P. & Li, K.C. (2008) Removal of arsenic from groundwater by electro-ultrafiltration. *Desalination*, 234 (1–3), 402–408.

Cheryan, M. (1998) *Ultrafiltration and microfiltration handbook*. Technomic Publishing Company, Lancaster, PA.

Christian, S.D., Tucker, E.E. & Scamehorn, J.F. (1995) Colloid-enhanced ultrafiltration in remediating wastewater and groundwater. *Speciality Chemicals*, 15 (3), 148–151.

Chwirka, J.D., Thomson, B.M. & Stomp, J.M. (2000) Removing arsenic from groundwater. *Journal American Water Works Association*, 92 (3), 79–88.

Chwirka, J.D., Colvin, C., Gomez, J.D. & Mueller, P.A. (2004) Arsenic removal from drinking water using the coagulation/microfiltration process. *Journal American Water Works Association*, 96 (3), 106–114.

De, S. & Mondal, S. (2012) *Micellar enhanced ultrafiltration*. CRC Press, Boca Raton, FL.

Drioli, E., Laganà, F., Criscuoli, A. & Barbieri, G. (1999) Integrated membrane operations in desalination processes. *Desalination*, 122 (2–3), 141–145.

Dunn, R.O., Scamehorn, J.F. & Christian, S.D. (1985) Use of micellar-enhanced ultrafiltration to remove dissolved organics from aqueous stream. *Separation Science and Technology*, 20 (4), 257–284.

Dunn, R.O., Scamehorn, J.F. & Christian, S.D. (1989) Simultaneous removal of dissolved organics and dissolved metal cations from water using micellar-enhanced ultrafiltration. *Colloids and Surfaces*, 35 (1), 49–56.

Ergican, E., Gecol, H. & Fuchs, A. (2005) The effect of co-occurring inorganic solutes on the removal of arsenic (V) from water using cationic surfactant micelles and an ultrafiltration membrane. *Desalination*, 181 (1–3), 9–26.

Fane, A.G. (1983) Factors affecting flux and rejection in ultrafiltration. *Separation and Purification Technology*, 4 (1), 15–23.

Fane, A.G. (1984) Ultrafiltration of suspensions. *Journal of Membrane Science*, 20 (3), 249–259.

Figoli, A., Cassano, A., Criscuoli, A., Islam Mozumder, M.S., Uddin, M.T., Islam, M.A. & Drioli, E. (2010) Influence of operating parameters on the arsenic removal by nanofiltration. *Water Research*, 44 (1), 97–104.

Gecol, H., Ergican, E. & Fuchs, A. (2004) Molecular level separation of arsenic (V) from water using cationic surfactant micelles and ultrafiltration membrane. *Journal of Membrane Science*, 124 (1), 105–119.

Geucke, T., Deowan, S.A., Hoinkis, J. & Pätzold, C. (2009) Performance of a small-scale RO desalinator for arsenic removal. *Desalination*, 239 (1–3), 198–206.

Ghosh, D., Medhi, C.R. & Purkait, M.K. (2011) Treatment of fluoride, iron and arsenic containing drinking water by electrocoagulation followed by microfiltration. *Proceedings of the 12th International Conference on Environmental Science and Technology, 8–10 September 2011, Rhodes island, Greece*. B-345.

Ghurye, G., Clifford, D. & Tripp, A. (2004) Iron coagulation and direct microfiltration to remove arsenic from groundwater. *Journal American Water Works Association*, 96 (4), 143–152.

Han, B.B., Runnells, T., Zimbron, J. & Wickramasinghe, R. (2002) Arsenic removal from drinking water by flocculation and microfiltration. *Desalination*, 145 (1–3), 293–298.

Ho, W. & Sirkar, K. (1992) *Membrane handbook*. van Mostrand Reinhold, New York, NY.

Huotari, H., Tragradh, G. & Huisman, I. (1999) Crossflow membrane filtration enhanced by an external DC electric field: a review. *Chemical Engineering Research and Design*, 77 (A5), 461–468.

Iqbal, J., Kim, H.J., Yang, J.S., Baek, K. & Yang, J.W. (2007) Removal of arsenic from groundwater by micellar-enhanced ultrafiltration (MEUF). *Chemosphere*, 66 (5), 970–976.

Jain, A. & Loeppert, R.H. (2000) Effect of competing anions on the adsorption of arsenate and arsenite by ferri-hydrite. *Journal of Environmental Quality*, 29 (5), 1422–1430.

Johnston, R., Heinjnen, H. & Wurzel, P. (2001) Safe water technology. In: Ahmed, M.F., Ali, M.A. & Adeel, Z. (eds.) *Technologies for arsenic removal from drinking water*. Matiar Manush, Dhaka, Bangladesh. pp. 1–98.

Juang, R.S., Xu, Y.Y. & Chen, C.L. (2003) Separation and removal of metal ions from dilute solutions using micellar-enhanced. *Journal of Membrane Science*, 218 (1–2), 257–267.

Kang, M., Kawasaki, M., Tamada, S., Kamei, T. & Magara, Y. (2000) Effect of pH on the removal of arsenic and antimony using reverse osmosis membranes. *Desalination*, 131 (1–3), 293–298.

Katsoyiannis, I.A. & Zouboulis, A.I. (2013) Removal of uranium from contaminated drinking water: a mini review of available treatment methods. *Desalination and Water Treatment*, 51 (13–15), 2915–2925.

Kim, D.H., Kim, K.W. & Cho, J. (2006) Removal and transport mechanisms of arsenics in UF and NF membrane processes. *Journal of Water and Health*, 4 (2), 215–223.

Kim, J. & Benjamin, M.M. (2004) Modeling a novel ion exchange process for arsenic and nitrate removal. *Water Research*, 38 (8), 2053–2062.

Košutić, K., Furač, L., Sipos, L. & Kunst, B. (2005) Removal of arsenic and pesticides from drinking water by nanofiltration membranes. *Separation and Purification Technology*, 42 (2), 137–144.

Liu, C.K. & Li, C.W. (2005) Combined electrolysis and micellar enhanced ultrafiltration (MEUF) process for metal removal. *Separation and Purification Technology*, 43 (1), 25–31.

Lohokare, H.R., Muthu, M.R., Agarwal, G.P. & Kharul, U. (2008) Effective arsenic removal using polyacrylonitrile-based ultrafiltration (UF) membrane. *Journal of Membrane Science*, 320 (1–2), 159–166.

Mathuthu, A. & Ephraim, J. (1993) Calcium-binding by fulvic-acids studied by an ion-selective electrode and an ultrafiltration method. *Talanta*, 40 (4), 521–526.

Meenakshi, A. & Maheshwari, R.C. (2006) Fluoride in drinking water and its removal. *Journal of Hazardous Materials*, 137 (1), 456–463.

Meng, X., Bang, S. & Korfiatis, G.P. (2000) Effects of silicate, sulfate, and carbonate on arsenic removal by ferric chloride. *Water Research*, 34 (4), 1255–1261.

Meng, X., Korfiatis, G.P., Christodoulatos, C. & Bang, S. (2001) Treatment of arsenic in Bangladesh well water using a household co-precipitation and filtration system. *Water Research*, 35 (12), 2805–2810.

Mohan, D. & Pittman, C.U. (2007) Arsenic removal from water/wastewater using adsorbents – a critical review. *Journal of Hazardous Materials*, 142 (1–2), 1–53.

Morel, G., Ouazzani, N., Graciaa, A. & Lachaise, J. (1997) Surfactant modified ultrafiltration for nitrate ion removal. *Journal of Membrane Science*, 134 (1), 47–57.

Mulder, M. (1998) *Basic principles of membrane technology*. Kluwer Academic Publisher, London, UK.

Ng, K.S., Ujang, Z. & Le-Clech, P. (2004) Arsenic removal technologies for drinking water treatment. *Reviews in Environmental Science and Bio/Technology*, 3 (1), 43–53.

Ning, R.Y. (2002) Arsenic removal by reverse osmosis. *Desalination*, 143 (3), 237–241.

Oh, J.I., Yamamoto, K., Kitawaki, H., Nakao, S., Sugawara, T., Rahman, M.M. & Rahman, M.H. (2000) Application of low-pressure nanofiltration coupled with a bicycle pump for the treatment of arsenic-contaminated groundwater. *Desalination*, 132 (1–3), 307–314.

Pookrod, P., Haller, K.J. & Scamehorn, J.F. (2004) Removal of arsenic anions from water using polyelectrolyte-enhanced ultrafiltration. *Separation Science and Technology*, 39 (4), 811–831.

Rahmanian, B., Pakizeh, M. & Maskooki, A. (2010) Micellar-enhanced ultrafiltration of zinc in synthetic wastewater using spiral-wound membrane. *Journal of Hazardous Materials*, 184 (1–3), 261–267.

Rautenbach, R. & Albrecht, R. (1986) *Membrane processes*. John Wiley & Sons, New York, NY.

Rivas, B.L., Maureira, A. & Geckeler, K.E. (2006a) Novel water-soluble acryloylmorpholine copolymers: synthesis, characterization, and metal ion binding properties. *Journal of Applied Polymer Science*, 101 (1), 180–185.

Rivas, B.L., Pooley, S.A., Pereira, E., Montoya, E. & Cid, R. (2006b) Poly(ethylene-alt-maleic acid) as complexing reagent to separate metal ions using membrane filtration. *Journal of Applied Polymer Science*, 101 (3), 2057–2061.

Rivas, B.L., Aguirre, M.C., Pereira, E., Moutet, J.C. & Saint-Aman, E. (2007) Capability of cationic water-soluble polymers in conjunction with ultrafiltration membranes to remove arsenate ions. *Polymer Engineering & Science*, 47 (8), 1256–1261.

Sabatè, J., Pujola, M. & Llorens, J. (2002) Comparison of polysulfone and ceramic membranes for the separation of phenol in micellar-enhanced ultrafiltration. *Journal of Colloid and Interface Science*, 246 (1), 157–163.

Saitúa, H., Campderrós, M., Cerutti, S. & Padilla, A.P. (2005) Effect of operating conditions in removal of arsenic from water by nanofiltration membrane. *Desalination*, 172 (2), 173–180.

Sato, Y., Kang, M., Kamei, T. & Magara, Y. (2002) Performance of nanofiltration for arsenic removal. *Water Research*, 36 (13), 3371–3377.

Saxena, A. & Shahi, V.K. (2007) pH controlled selective transport of proteins through charged ultrafilter membranes under coupled driving forces: an efficient process for protein separation. *Journal of Membrane Science*, 299 (1–2), 211–221.

Saxena, A., Tripathi, B.P., Kumar, M. & Shahi, V.K. (2009) Membrane-based techniques for the separation and purification of proteins: an overview. *Advances in Colloid and Interface Science*, 145 (1–2), 1–22.

Seidel, A., Waypa, J.J. & Elimelech, M. (2001) Role of charge (Donnan) exclusion in removal of arsenic from water by a negatively charged porous nanofiltration membrane. *Environmental Engineering Science*, 18 (2), 105–113.

Shih, M.C. (2005) An overview of arsenic removal by pressure-drivenmembrane processes. *Desalination*, 172 (1), 85–97.

Song, S., Lopez-Valdivieso, A., Hernandez-Campos, D.J., Peng, C., Monroy-Fernandez, M.G. & Razo-Soto, I. (2006) Arsenic removal from high-arsenic water by enhanced coagulation with ferric ions and coarse calcite. *Water Research*, 40 (2), 364–372.

Sung, M., Huang, C.P., Weng, Y.H., Lin, Y.T. & Li, K.C. (2007) Enhancing the separation of nano-sized particles in low-salt suspensions by electrically assisted cross-flow filtration. *Separation and Purification Technology*, 54 (2), 170–177.

Tabatabai, A., Scamehorn, J.F. & Christian, S.D. (1995a) Water softening using polyelectrolyte-enhanced UF. *Separation Science and Technology*, 20 (2), 211–224.

Tabatabai, A., Scamehorn, J.F. & Christian, S.D. (1995b) Economic feasibility study of polyelectrolyte-enhanced ultrafiltration (PEUF) for water softening. *Journal of Membrane Science*, 100 (3), 193–207.

Tangvijitsri, S., Saiwan, C., Soponvuttikul, C. & Scamehorn, J.F. (2002) Polyelectrolyte-enhanced UF of chromate, sulfate, and nitrate. *Separation Science and Technology*, 37 (5), 993–1007.

Uddin, M.T., Mozumder, M.S.I., Figoli, A., Islam, M.A. & Drioli, E. (2007a) Arsenic removal by conventional and membrane technology: an overview. *Indian Journal of Chemical Technology*, 14 (5), 441–450.

Uddin, M.T., Mozumder, M.S.I., Islam, M.A., Deowan, S.A. & Hoinkis, J. (2007b) Nanofiltration membrane process for the removal of arsenic from drinking water. *Chemical Engineering & Technology*, 30 (9), 1248–1254.

Urase, T., Oh, J. & Yamamoto, K. (1998) Effect of pH rejection of different species of arsenic by nanofiltration. *Desalination*, 117 (1–3), 11–18.

Vrijenhoek, E.M. & Waypa, J.J. (2000) Arsenic removal from drinking water by a loose nanofiltration membrane. *Desalination*, 130 (3), 265–277.

Weber, K. & Stahl, W. (2003) Determination of the zeta potential of a filter cake by means of permeation experiments. *Chemical Engineering & Technology*, 26 (5), 549–553.

Weng, Y.H., Chaung-Hsieh, L.H., Lee, H.H., Li, K.C. & Huang, C.P. (2005) Removal of arsenic and humic substances (HSs) by electro-ultrafiltration (EUF). *Journal of Hazardous Materials*, 122 (1–2), 171–176.

Wickramasinghe, S.R., Han, B.B., Zimbron, J., Shen, Z. & Karim, M.N. (2004) Arsenic removal by coagulation and filtration: comparison of groundwaters from the United States and Bangladesh. *Desalination*, 169 (3), 231–244.

Zaw, M. & Emett, M.Y. (2002) Arsenic removal from water using advanced oxidation processes. *Toxicology Letters*, 133 (1), 113–118.

Zhou, D., Zhao, H., Price, W.E. & Wallace, G.G. (1995) Electrochemically controlled transport in a dual conducting polymer membrane system. *Journal of Membrane Science*, 98 (1–2), 173–176.

CHAPTER 3

Fluoride and uranium removal by nanofiltration

Stefan-André Schmidt, Tiziana Marino, Catherine Aresipathi, Shamim-Ahmed Deowan, Priyanath N. Pathak, Prasanta Kumar Mohaptra, Jan Hoinkis & Alberto Figoli

3.1 INTRODUCTION

Membrane separation processes have been extensively used for the fractionation/concentration of suspended particles and dissolved substances in different streams. In this context, nanofiltration (NF) appears attractive and has shown potential for the treatment of wastewaters. Reverse osmosis (RO), currently in widespread use to produce drinking or irrigation water from briny waters or seawater (Hassan *et al.*, 1998) seriously challenges the distillation process, whereas ultrafiltration (UF) constitutes a valuable aid for the fractionation and concentration of colloidal substances contained in seawater (Teuler *et al.*, 1999). Based on both the size of separated species and pressures involved, NF is in between the UF and RO and it has been more recently developed than the other two processes. The NF process is characterized by low energy consumption, and is usually applied to separate multivalent ions from monovalent ones, and also for separation between ions with the same valence. NF membranes are designed as "loose" or "low pressure" RO membranes; these terms define a first clear objective to NF membranes: partial rejection of salts (Forare, 2009). The main application of NF is in the wastewater treatment, water softening, desalination of brackish water and removal of several compounds, such as pollutants, pesticides, dyes, and organic solvents, from effluents. NF is a transition zone, also in terms of physicochemical interactions and transport mechanisms (solution-diffusion mechanism), with features between UF and RO (Li *et al.*, 2008). The membranes for NF have slightly larger pore size than RO membranes. In fact, NF membranes have a molecular weight cutoff (MWCO) to dissolved organic solutes of 200–1000 Dalton; while, for RO membranes, the molecular weight cut-off is less than 50 Dalton. NF membrane pore size is in the order of nanometers, particularly $\leq$2 nm, and the driving force used is a pressure in the range of 1.0–2.5 MPa (Sheppard *et al.*, 2005). These membranes are usually asymmetric, with negative surface charge at neutral and alkaline drinking water pH; which explains higher rejection of cations and anions. For this reason, the rejection (or permeation) of salts depends not only on the membrane characteristic and molecular size, but also Donnan exclusion effects. These characteristics lead to having similar ion separation degrees to RO membranes but with higher water fluxes (Forare, 2009). The most important works, dealing with the removal of uranium (U) and fluoride (F^-) by NF mainly from water, are discussed and reported in this section.

3.2 COMMON REMOVAL TECHNOLOGIES

3.2.1 *Fluoride*

Different water defluorination methods have been investigated: adsorption (Ben Nasr *et al.*, 2011; Fan *et al.*, 2003), ion exchange (Meenakshi *et al.*, 2007; Solangi *et al.*, 2009), chemicals addition to cause precipitation (Nath *et al.*, 2010; Turner *et al.*, 2005) and membrane processes such as RO, electrodialysis (ED) and NF (Amor *et al.*, 2001; Ben Nasr *et al.*, 2011; Tahaikt *et al.*, 2007). RO, activated alumina and ED are the most common F^- removal technologies (Adhikary *et al.*, 1989; Amor *et al.*, 2001; Min *et al.*, 1984; Cohen *et al.*, 1998; WHO, 2006). Activated alumina is one of a variety of precipitation and adsorption materials. Adsorption materials are widely used for F^- treatment in groundwater due to their simple handling and low costs. In Senegal, Diawara *et al.* (2011) described charcoals and clays as being efficient in F^- removal. The performance of clay in absorbing fluorine ions is strongly bounded to its specific area, the quality of the raw water, (physically and chemically) and the geometry of the filtration module. Gumbo *et al.* (1995) reported findings of a pilot defluorination plant installed in Tanzania. They described the investigations of the efficiency for the Magnesite and Nalgonda technique. For the Nalgonda technique high concentrations of aluminum and lime are used, while for the Magnesite technique, magnesia obtained by calcination of Magnesite, have been investigated. A ratio of 800 mg L^{-1} alum and 80 mg L^{-1} lime reduced the F^- concentration in average from 22 to 3.5 mg L^{-1}. After passing a filter-bed filled with calcinated magnesite the F^- concentration decreased by another 1 mg L^{-1} but the pH raised to around 10 and required further treatment (Gumbo *et al.*, 1995). As another adsorbent, developed by Chen *et al.* (2012), Fe-Ti has been tested and considered as efficient and economical in terms of F^- removal from drinking water. It has been reported that this adsorbent is much cheaper and has a higher adsorption capacity (about 47 mg g^{-1}) than other adsorbing materials, like zirconium oxide and rare earth metal oxides. Vaishali *et al.* (2013) described the adsorption capacity of *Citrus limonum* leaf. His group found a maximum defluoridation capacity of 70% of 2 mg L^{-1} F^- ion. The capacity was found to be dependent on the pH, adsorbent dose, contact time and initial F^- concentrations.

Despite its simplicity, the use of adsorption media has major drawbacks in particular for application in remote rural areas which is of high importance for worldwide groundwater treatment: (i) supply of adsorption media needs to be ensured, (ii) safe disposal of exhausted media or regeneration, (iii) adsorption capacity largely depends on the particular feed water quality such as e.g. presence co-ions, pH, hence in many cases safe compliance with MCL cannot be ensured.

3.2.2 *Uranium*

Solid phase/solvent extraction and membrane separation are efficient technologies for the removal of U from water (Tang *et al.*, 2003). Solid phase extraction is a highly efficient methodology for the preconcentration and purification of U traces from water (Kantipuly *et al.*, 1990). This technique is based on the use of chelating resin materials, such as modified cellulose/silica, activated carbon/alumina and polymeric resins (Hancock and Martell, 1989; Helfferich, 1962; Mahmoud and Al Saadi, 2001; Schmitt and Pietrzyk, 1985; Yaman and Gucer, 1995; Yamini *et al.*, 2002) featuring anionic functional groups (e.g., strong acid-SO_3H; weak acid-COOH; strong base-NR_3Cl; weak base-NH_2RCl) able to bind U to the resins (Kabay and Egawa, 1993; Rivas *et al.*, 2003). Fibrous polymeric adsorbents, containing amidoxime groups, have shown a growing attention in view of their several advantages, such as, i.e., high adsorption ability and selectivity for the uranyl ions uptake and easy handling and eco-compatibility (Abbasi and Streat, 1994). Chattanathan *et al.* (2013), investigated the effect of hydroxyapatite prepared from catfish bones in the removal of U from wastewater, achieving promising results in terms of adsorption efficiency. Qadeer and Hanif (1994) studied the influence of kinetic factors on the U ions adsorption on activated charcoal from aqueous solutions. Kutahyali and Eral (2004) reported U adsorption experiments on chemically activated carbon. Solvent extraction is generally based on the use of a pregnant leach solution which provides the chelation or ion-association of U and the subsequent extraction of uraniferous ions in a suitable solvent (Khopkar and Charmers, 1970; Ritcey and Ashbrook, 1984). In the leaching step, uranyl sulfates or carbonates are formed and can be efficiently extracted in anionic solvents. Di-2-ethylhexyl phosphoric acid/trioctylphosphine oxide is a good mixture for the extraction of U from phosphoric acid solutions (Mohsen *et al.*, 2013). Koban and Bernhard (2004) proposed the use of glycerol-1-phosphate for the formation of U^{6+} complexes. Singh *et al.* (2001) studied the extraction power of the system di-nonyl phenyl phosphoric acid, with di-butyl butyl phosphonate in an aliphatic diluent, obtaining promising results for the U stripping from concentrated phosphoric acid solutions. Kulkarni (2003) used trioctylphosphine oxide in paraffin oil and sodium carbonate as carrier and stripping agent, respectively for the purification of acidic wastes. One of the most promising technologies for water treatment is membrane separation. Villalobos-Rodriguez *et al.* (2012) investigated the UF removal of U using composite activated carbon cellulose triacetate membranes. The permeation based on the molecular sieving mechanism coupled with the adsorption of the cationic species by the carbon particles improved the U rejection from the aqueous medium.

3.3 REMOVAL OF DISSOLVED FLUORIDE AND URANIUM BY NF

3.3.1 *Fluoride*

NF offers selective desalination and is generally used to remove divalent ions such as sulfate and calcium ions and also to separate ions of the same valency (Ben Nasr *et al.*, 2011; van der Bruggen *et al.*, 2004). Therefore NF is typically not applied for removal of monovalent ions such as F^-. However, in the case of F^- due to its small ionic radius compared to e.g. chloride, and a consequently larger hydrate shell, the NF can offer a viable option. Hence, the smaller the ionic radii, the higher the hydration numbers; the larger the hydrated radii, the more difficult its transfer across the membrane will be (Paugham *et al.*, 2004). The NF operates under significantly lower pressure. As the treatment of F^- contaminated groundwater by NF requires only a comparably low pressure of about 0.5–1.0 MPa, this technique is a very energy efficient solution in comparison to RO. It can be easily operated by renewable energies like photovoltaic or wind turbines and therefore has a great potential to lower capital and operating costs. In fact, for applications in rural areas of less developed countries, the basic requirements are: low cost, very simple to operate and easy to maintain water treatment. Additionally, water remineralization with NF is not needed (Ben Nasr *et al.*, 2011). However, up to now, studies on F^- removal using NF membranes are limited to laboratory or small pilot trials. Table 3.1 gives an overview of different rejection rates for F^- removal. As shown in this table, NF membranes can remove up to 99% of F^- from water. Lhassani *et al.* (2001) studied defluorination of brackish water model solutions with halides such as Cl^- and I^- by use of the Dow NF 70. It has been shown that since F^- ions are more solvated than chloride and iodide ions, the small F^- ions can be better retained than the bigger halides. Under reduced pressure, this effect is even higher when selectivity is the most. By adjusting the operating conditions, the ions of same valency can be removed selectively. Hence, Diawara *et al.* (2003) also analyzed the F^- removal from brackish water model solutions by use of a spiral wound NF 45-2540. The focus of this work was to test the rejection of fluoride co-existing with other sodium and lithium halides. It has been shown that a selective defluorination can be achieved by the use of this membrane. The NF 70 membranes, that have been investigated by Pontie *et al.* (2002), showed an increasing retention $F^- > Cl^- > Br^-$ at low pressure following the increasing hydrated ionic radius. Choi *et al.* (2001) have investigated two commercial NF membranes from Nitto Denko Corporation regarding the influence of the co-existing ions on the F^- rejection. The average F^- rejection found to be between 70–72%. The performance of three thin-film composite NF membranes have been studied with model solutions, for the F^- removal, under different operating conditions, concentration up to 1000 mg L^{-1} at costant pressure of about 1.37 MPa (Chakrabortty *et al.*, 2013). Additionally, in this study a mathematical model has been used to interpret the performance of the NF membranes at different F^- contents. By using this model, the membrane performance has been calculated based on three membrane parameters. In some cases, the F^- rejection was higher than 80%. The F^- removal by NF on F^--spiked groundwater in Morocco has been studied by Tahaikt *et al.* (2007; 2008). Four different spiral-wound NF 4040 membranes have been tested.

By using F^--spiked groundwater (1.8–22 mg L^{-1}) the F^- rejection of the NF 90 was highest (96–99%) and having the best performance among the tested membranes. As the retention of the NF 270, NF 400 and TR 60 was in a similar range (about 50–88%) they might be applied in treating groundwater with lower initial F^- content. Hoinkis *et al.* (2011) showed that the fluoride level in permeate was below the MCL up to 10 mg L^{-1} in feed for the NF 270 and

Table 3.1. NF membranes rejection [%] for fluoride removal.

Membrane and manufacturer	Water origin	Rejection [%]	Permeate flux [L h^{-1} m^{-2}]	Reference
NTR 7250, Nitto Denko Corporation	Model solutions with Na^+, Ca^{2+}, Mg^{2+}, SO_4^{2-}, Cl^-	70	–	Choi *et al.* (2001)
NTR 7450, Nitto Denko Corporation		72		
NF 70-2540 spiral wound, Dow/Filmtec	Model solutions with Na^+, Cl^-, I^-	Single solute >90 at low recovery	20–25 (at 1.0 MPa)	Lhassani *et al.* (2001)
NF 55, Dow Filmtec	Model solutions with K^+, F^-, Cl^- and Br^-	95	–	Pontie *et al.* (2002)
NF 70, Dow Filmtec		55		
NF 90, Dow Filmtec		n/a		
NF 45-2540 spiral wound, Dow Filmtec	Model solutions with Na^+ and Li^+, Cl^-, I^-	NaF: 91–96 LiF: 88–93	–	Diawara *et al.* (2003)
SR-1, Koch	Model solutions with Na^+	0–70	60–80	Hu *et al.* (2006)
DS-5-DL, Osmonics		10–80	50–65	
HS-51-HL, Osmonics		20–95	58–80	
NF 90-4040 spiral wound, Dow Filmtec	Fluoride spiked natural groundwater	96–99	46 (at 1.0 MPa)	Tahaikt *et al.* (2007)
NF 400-4040 spiral wound, Dow Filmtec		50–86	49 (at 1.0 MPa)	
NF 90-4040, Dow Filmtec	Fluoride spiked natural groundwater	NF 90: 98–90	59 (at 1.2 MPa)* 64 (at 0.9 MPa)*	Tahaikt *et al.* (2008)
NF 270-4040, Dow Filmtec		83–88	61 (at 0.5 MPa)*	
TR60-4040, Toray		74–86	64 (at 0.9 MPa)*	
NF 270, Dow Filmtec	Fluorine brackish groundwater	63.3–71	78 (at 0.79–0.89 MPa)	Diawara *et al.* (2011)
NF 270, Dow Filmtec	Model fluoride water (20 mg/L, 25°C, pH 7)	87–88	60 (at 0.5 MPa), 112 (at 1 MPa) 142 (at 1.5 MPa)	Hoinkis *et al.*. (2011)
NF 90, Dow Filmtec		98–98.6	30 (at 0.5 MPa), 62 (at 1 MPa), 88 (at 1.5 MPa), 116 (at 2 MPa)	
NF 5 Applied Membranes Inc.	Groundwater	57	120 (at 1.0 MPa, 25°C)	Ben Nasr *et al.* (2013)
NF 9 Applied Membranes Inc.		88	85 (at 10 MPa, 25°C)	
NF-1 Sepro Membranes Inc. (USA)	Fluorine brackish groundwater from Asanjola village, West Bengal, India	98.5	158 (at 1.373 MPa)	Chakrabortty *et al.* (2013)
NF-2 Sepro Membranes Inc. (USA)		91	214 (at 1.373 MPa)	
NF-20 Sepro Membranes Inc. (USA)		n/a	365 (at 1.373 MPa)	

up to 20 mg L^{-1} for the NF 90, respectively. In addition, they could show no significant influence of HCO_3^- to the fluoride rejection, whereas at the pH of 5 the fluoride rejection was noticeably lower. Diawara *et al.* (2011) published their outcomes of a pilot unit installed in Senegal for the treatment of fluorine rich brackish water. The module of the pilot unit is composed of 169 circular flat membranes with a total area of 7.605 m^2. The permeate flow of 10 L min^{-1} has been kept constant by adjusting the feed pressure between 0.79 and 0.89 MPa. However, after 300 hours of running, a chemical cleaning was necessary due to fouling on the membrane surface. Ben Nasr *et al.* (2013) conducted the F^- removal of two commercial NF membranes NF5 and NF9. After tests with model water a pilot unit with underground water was studied. The difference in both membranes lies mainly in different salt rejection rates. Beside the initial F^- content, the effect of chloride, sulfate and calcium were analyzed since these ions usually co-exist in groundwater. It has been shown that the NF membranes reject divalent anions very strongly and in addition, the smaller the monovalent ion the better it is retained. They described that the reason for that is derived from the solvation energy of the ions by water. Since chloride ions are less solvated they are lower retained than F^- ions (Ben Nasr *et al.*, 2013). Chakrabortty *et al.* (2013) measured an increase in flux by rising transmembrane pressure. This effect is well reported in literature (e.g., Tahaikt *et al.*, 2007). By comparing the transmembrane pressure with the volumetric flux an optimum pressure of 14 kg cm^{-2} (1.373 MPa) was considered because beyond that no further improvement in F^- rejection was measured. An increase in flux came along with the increase of the cross flow rate and indicated a strong correlation between both parameters. An economic evaluation for a plant of a capacity of 10 m^3 d^{-1} indicated costs of 1.17 US\$ m^{-3} including capital costs and operating costs (Chakrabortty *et al.*, 2013).

3.3.2 *Uranium*

The removal of U from drinking water was evaluated by Raff *et al.* (2009). These authors studied the influence of the pH on the membrane uranyl rejection in deionized water. In particular, the performance of five types of NF membranes, of which three purchased from the Companies Osmonics Desal (Desal 5DK, Desal 5DL and Desal 51HL) and two from Dow (NF90 and NF45), were investigated. In each case, the U rejection was between 81 and 99%. The best results were observed using the membranes Desal 5DK (96–99%) and Desal 5DL (94–98%). These promising results indicated that the rejection of two negatively charged uranyl carbonate complexes $UO_2(CO_3)_2^{2-}$ and $UO_2(CO_3)_3^{4-}$ was 94% or greater. Frave-Reguillon *et al.* (2003) investigated the removal of U dissolved in seawater. A simulated seawater was prepared by mixing uranyl nitrate $UO_2(NO_3)_2$, distilled water, Na_2CO_3 and different ions such as Na^+, K^+, Ca^{2+} and Mg^{2+}. For this study, four different NF flat sheet membranes purchased from Osmonics and having different molecular weight cut off (G50:8000 Da, G20:3500 Da, G10:2500 Da and 5DL:150–300 Da), were used. When the pH of the aqueous solution was 8.3, $UO_2(CO_3)_2^{2-}$ and $UO_2(CO_3)_3^{4-}$ were the most abundant species. Initially, these authors studied the U retention using a feed solution containing U carbonate with sodium chloride. The U rejection decreased with the increasing of NaCl concentration. Only for the G50 membrane was a rapid decrease at already low concentrations of NaCl (1 g L^{-1}) observed. The other three membranes (5DL, G10 and G20) showed a retention coefficient greater than 70%. The tests carried out with the simulated seawater, the U^{6+} retention coefficient (R) decreased to 50% for the G20 membranes. The retention coefficient (R%) is defined as $[(1 - C_p/C_f) \times 100]$, where C_p and C_f are concentrations in mol L^{-1} of the metal ions in permeate and the feed solutions, respectively. For G10 and 5DL there was no significant decrease in terms of U^{6+} retention, but for 5DL the selectivity for U^{6+}/Ca^{++} was dramatically low. Only the G10 membrane showed a high retention coefficient for U^{6+} and a low retention coefficient for sodium and calcium, leading to high U^{6+}/Ca^{++} and U^{6+}/Na^+ selectivities. The best efficiency was obtained using the G10 membrane, which evidenced the possibility to selectively filter U. The removal of U^{6+}, dissolved in drinking water, was also investigated by Favre-Reguillon *et al.* (2008). These authors demonstrated that NF membranes were able to reject U^{6+} from mineral water with a relatively high selectivity, despite a high concentration of alkaline and alkaline-earth cations. They studied the performance of three different commercial polyamides supported on a polysulfone NF membranes (Osmonics). These membranes presented different MWCO and isoelectric point (pI). G10 (2500 Da, 3.7 pI), DL (150–300 Da, –), DK (150–300 Da, 4.0 pI). They studied the solutions deriving from the addition of uranyl nitrate ($UO_2(NO_3)_2 \cdot 6H_2O$) to three different commercial mineral waters. The rejection of U^{6+} was 40%, 95% and 99% for the G10, DL and DK membranes, respectively; the highest cut-off membrane exhibited the lowest solute rejection for the same water composition. The rejection of alkaline and alkaline-earth ions marked the same trend for all the tested membranes. A positive trend was observed for the DL membrane. In fact, in all cases U^{6+} rejection was high (95%), whereas that for the monovalent and divalent ions was low. On the contrary, the DK membrane demonstrated the ability to highly reject not only the U^{6+}, but also the monovalent and divalent ions. Moreover, the U^{6+} rejection obtained using the G10 membrane was the lowest. Kryvoruchko *et al.* (2013) and Yurlova (2010) studied the purification of U-containing water by NF using the OPMN-P, a polyamide commercial membrane (Vladipor Company). The innovation of this work resides in the simultaneous use of modified montmorillonite (Cherkassy Deposit) with polyethyleneimine (MW, 2000), in NF process. This study demonstrated the possibility to couple the use of montmorrillonite as a sorbent for water purification from U^{6+} and the NF separation process. The application of modified montmorillonite in NF process allowed to achieve U retention coefficients as high as 0.999. In another study, U^{6+} removal by micellar-enhanced ultra- and nano- filtration was investigated to study the influence of the concentration and the steric structure of the surfactants (sodium dodecyl sulfate and sodium dodecyl benzene sulfonate), and the pH. The steric hindrance of the examined surfactants affected the efficiency of treatment of U-contaminated waters. Interestingly, water decontamination from U^{6+} was the most efficient when the surfactants were added in concentrations close to their critical micelle concentrations. Richards *et al.* (2011) evaluated the effects of fluctuating energy and pH on the retention of dissolved contaminants from real Australian groundwaters, using a solar (photovoltaic) powered UF-NF/RO system using four NF/RO membranes (BW30, ESPA4, NF90, and TFC-S). Whereas the fluctuations in energy affected pressure and flow, the solar irradiance levels influenced the retention of fluoride, magnesium, nitrate, potassium, and sodium. On the other hand, the retention of calcium, strontium, and U was found to be very high and independent of the solar irradiance, which was a combined effect of size and charge exclusion. The groundwater characteristics also affected the retention and, therefore, solutes were categorized into two groups according to their retention behavior as a function of pH: (1) pH-independent retention (arsenic, calcium, chloride, nitrate, potassium, selenium, sodium, strontium, and sulfate) and (2) pH-dependent retention (copper, magnesium, manganese, molybdenum, nickel, uranium, vanadium and zinc). The retention of Group 1 solutes was typically high and attributed to steric effects. Group 2 solutes had dominant, insoluble species under specific conditions which led to deposition on the membrane surface (and thus varying apparent retention). However, the renewable energy membrane system was effective in removing a large number of groundwater solutes for a range of real energy and pH conditions. Oliveira *et al.* (2012; 2013) evaluated the performance of a NF membrane for treatment of a low-level radioactive liquid waste (carbonated water during conversion of UF_6 to UO_2) through static and dynamic tests. Membrane hydraulic permeability, permeate flow and selectivity were measured before and after its immersion in the liquid waste (static tests). In the dynamic tests, to determine the performance of NF membranes for U removal, the waste was permeated through the membrane at 0.5 MPa. The surface layer of the membrane was characterized by zeta potential, field emission microscopy, atomic force spectroscopy and infrared spectroscopy. The static test showed that the membrane surface charge was not significantly changed; the U rejection after the dynamic test was 99%. Prabhakar *et al.* (1996) compared the performance of cellulose acetate (CA) and polyamide (PA) membranes for the removal of radioactive species from ammonium diuranate filtrate effluents (ADUF). Even though the CA based membrane was promising in terms of very good decontamination factors (DF) and volume reduction factors (VRF). However, it was associated with the limitation of short membrane life which put was against its large scale application for the intended purpose. Therefore, PA based RO, NF, and UF membranes were tested on real effluents corresponding to specific activity levels of microcuries/liter containing ~4% ammonium nitrate. These treatments yielded decontaminated streams containing nanocurie/liter levels of radiocontaminants both for RO and NF membranes. In addition, the NF membranes showed the potential to achieve very high VRF and better decontamination factors owing to their poor ammonium nitrate rejection characteristics and the consequent maintenance of permeate fluxes. The studies indicate the viability of the NF process for the treatment of ammonium diuranate filtrate effluents in large scale. NF has the advantage of very low solute rejection for monovalent species, probably due to their very small sizes (hydrated radii), and higher rejection for multivalent species which are large enough (i.e., more than the critical pore diameter) for rejections based on physicochemical interactions. Therefore, solute rejection of ammonium nitrate was very low and the DF values were very high. Rana *et al.* (2013) have recently reviewed the efficacy of radioactive decontamination by membrane processes. These authors have classified the membrane technology into different processes and, for each process, stated progresses made since the onset of this millennium in the radioactive decontamination of water. The new directions are shown considering the growth made in membrane manufacturing and membrane processes. The combined efforts of researchers engaged in membrane and membrane process design with those engaged in nuclear waste treatment near the plant sites were highlighted. Table 3.2 summarizes the experimental results reported in the previously presented

Table 3.2. NF membranes rejection [%] for uranium removal.

Membrane and manufacturer	Water origin	Rejection [%]	Permeate flux [$L\,h^{-1}\,m^{-2}$]	Reference
Desal 5 DK, Osmonics Desa Companies	Model solutions with $NaHCO_3$ (pH 7.3–8.3)	96–99	3.5–7 (at 0.8 MPa)	Raff *et al.* (1999)
Desal 5 DL, Osmonics Desa Companies	Model solutions with $NaHCO_3$	94–98	n/a	
Desal 51 HL, Osmonics Desa Companies	De-ionized water at pH 6.7	88	n/a	
NF 90, Dow Chemical	De-ionized water at pH 6.7	82	n/a	
NF 45, Dow Chemical	De-ionized water with $NaHCO_3$ at pH 5.9	81	n/a	
G50, Osmonics	Model seawater solution	n/a	n/a	Frave-Reguillon *et al.* (2003)
G20, Osmonics	Model seawater solution	~50	n/a	
G10, Osmonics	Model seawater solution	~83	n/a	
5DL, Osmonics	Model seawater solution	~95	n/a	
G10, Osmonics	Commercial mineral water	40	n/a	Frave-Reguillon *et al.* (2008)
DL, Osmonics	Commercial mineral water	95	n/a	
DK, Osmonics	Commercial mineral water	99	n/a	
OPMN_P, Vladipor Company	Distilled water	99	n/a	Yurlova *et al.* (2010)
BW30, Koch	Groundwater	~100	n/a	Richards *et al.* (2011)
SW-0, DOW	Radioactive waste solution at pH 9.4	94	2.4	Oliveria *et al.* (2013)
Cellulose acetate membrane, HMIL	Ammonium diuranate filtrate effluents (ADUF)	n/a	n/a	Prabhakar *et al.* (1996)
Polyamide membrane, HMIL	Ammonium diuranate filtrate effluents (ADUF)	n/a	n/a	

works. As evidenced by the rejection results, NF membranes represent a very efficient system to remove the dissolved U.

3.4 CONCLUSIONS AND OUTLOOK

NF membrane separation offers the possibility to replace the common water treatment processes by a single-step procedure with low operating and energy costs. Several successful works on the removal of dissolved F^- and U from water by NF have demonstrated the opportunity to meet regulations for their lowered content in groundwater. In particular, it has been shown that even at elevated influent levels, NF membranes have the ability to reject F^- below the maximum contaminant level (MCL) of $1.5\,mg\,L^{-1}$ recommended by WHO (WHO, 2014). Also for the U removal, it has been evidenced as commercial NF membranes are able to reject U^{6+} from mineral water with high selectivity (Frave-Reguillon *et al.*, 2008), making this process suitable in the reduction of U content in drinking water to less than the WHO maximum admissible concentration ($0.002\,mg\,L^{-1}$). Nevertheless, some drawbacks, such as the fouling, the cost and the maintenance of the membranes, still limit the use of NF separation processes on an industrial scale. Pre-treatment processes may be suitable to largely eliminate pollutants from groundwaters such as viruses and bacteria and to prevent the membrane fouling. In this context, implementing a hybrid system, e.g. coupling NF with adsorption, RO or UF, may greatly enhance the efficiency of the removal process. Based on present investigation, it can be concluded that further studies on the concentration fraction is the key parameter to improve F^- and U removal by NF.

REFERENCES

Abbasi, W.A. & Streat, M. (1994) Adsorption of uranium from aqueous solutions using activated carbon. *Separation Science and Technology*, 29, 1217–1230.

Adhikary, S., Tipnis, U., Harkare, W. & Govindan, K. (1989) Defluoridation during desalination of brackish water by electrodialysis. *Desalination*, 71, 301–312.

Amor, Z., Bariou, B., Mameri, N., Taky, M., Nicolas, S. & Elmidaoui, A. (2001) Fluoride removal from brackish water by electrodialysis. *Desalination*, 133, 215–223.

Ben Nasr, A., Walha, K., Charcosset, C. & Ben Amar, R. (2011) Removal of fluoride ions using cuttlefish bones. *Journal of Fluorine Chemistry*, 132, 57–62.

Ben Nasr, A., Charcosset, C., Ben Amar, R. & Walha, K. (2013) Defluoridation of water by nanofiltration. *Journal of Fluorine Chemistry*, 150, 92–97.

Bugaisa, S.L. (1971) Significance of fluorine in Tanzania drinking water. In: *Proceedings Conference Rural Water Supply in East Africa, 5–8 April 1971, Dar-Es-Salaam, Tanzania*. pp. 107–113.

Chakrabortty, S., Roy, M. & Pal, P. (2013) Removal of fluoride from contaminated groundwater by cross flow nanofiltration: transport modeling and economic evaluation. *Desalination*, 313, 115–124.

Chattanathan, S.A., Clement, T.P., Kanel, S.R., Barnett, M.O. & Chatakondi, N. (2013) Remediation of uranium-contaminated groundwater by sorption onto hydroxyapatite derived from catfish bones. *Water, Air, & Soil Pollution*, 224, 1–9.

Chen, L., He, B.-Y., He, S., Wang, T.-J., Su, C.-L. & Jin, Y. (2012) Fe-Ti oxide nano-adsorbent synthesized by co-precipitation for fluoride removal from drinking water and its adsorption mechanism. *Powder Technology*, 227, 3–8.

Choi, S., Yun, Z., Hong, S. & Ahn, K. (2001) The effect of co-existing ions and surface characteristics of nanomembranes on the removal of nitrate and fluoride. *Desalination*, 133, 53–64.

Cohen, D. & Conrad, H. (1998) 65,000 GPD fluoride removal membrane system in Lakeland, California, USA. *Desalination*, 117, 19–35.

Diawara, C.F., Lo, S.M., Rumeau, M., Pontie, M. & Sarr, O. (2003) A phenomenological mass transfer approach in nanofiltration of halide ions for a selective defluorination of brackish drinking water. *Journal of Membrane Science*, 219, 103–112.

Diawara, C.K., Diop, S.N., Diallo, M.A., Farcy, M. & Deratani, A. (2011) Performance of nanofiltration (NF) and low pressure reverse osmosis (LPRO) membranes in the removal of fluorine and salinity from brackish drinking water. *Journal of Water Resource and Protection*, 3, 912–917.

Environmental Data Explorer, compiled from IGRAC. Available from: http://geodata.grip.unep.ch [accessed February 2014].

Fan, X., Parker, D.J. & Smith, M.D. (2003) Adsorption kinetics of fluoride on low cost materials. *Water Research*, 37, 4929–4937.

Favre-Réguillon, A., Lebuzit, G., Foos, J., Guy, A., Draye M. & Lemaire, M. (2003) Selective concentration of uranium from seawater by nanofiltration. *Industrial & Engineering Chemistry Research*, 42, 5900–5904.

Favre-Réguillon, A., Lebuzit, G., Murat, D., Foos, J., Mansour, C. & Draye, M. (2008) Selective removal of dissolved uranium in drinking water by nanofiltration. *Water Research*, 42, 1160–1166.

Forare, J. (ed.) (2009) Drinking water: sources sanitation and safeguarding. The Swedish Research Council Formas, Stockholm, Sweden. p. 74.

Grenthe, I., Fuger, J., Konings, R.J.M., Lemire, R.J., Muller, A.B., Nguyen-Trung, C. & Wanner, H. (1992) *Chemical thermodynamics of uranium*. North-Holland, Amsterdam, The Netherlands.

Gumbo, F.J. & Mkongo, G. (1995) Water defluoridation for rural fluoride affected communities in Tanzania. In: Dahi, E. & Bregnhøj, H. *1st International Workshop on Fluorosis Prevention and Defluorination of Water, 18–22 October 1995 Ngurdoto, Tanzania*. pp. 109–114.

Hancock, R.D. & Martell, A.E. (1989) Ligand design for selective complexation of metal ions in aqueous solution. *Chemical Reviews*, 89, 1875–1882.

Hassan, A.M., Al-Sofi, M.A.K., Al-Amoudi, A.S., Jamaluddin, A.T.M., Farooque, A.M., Rowaili, A., Dalvi, A.G.I., Kither, N.M., Mustafa, G.M. & Al-Tisan, I.A.R. (1998) A new approach to membrane and thermal seawater desalination processes using nanofiltration membranes (Part 1). *Desalination*, 118, 35–51.

Helfferich, F. (1962) *Ion exchange*. McGraw-Hill Book Company, New York, NY.

Hoinkis, J., Valero-Freitag, S., Caporgno, M.P. & Pätzold C. (2011) Removal of nitrate and fluoride by nanofiltration – a comparative study. *Desalination and Water Treatment*, 30, 278–288.

Hu, K. & Dickson, J.M. (2006) Nanofiltration membrane performance on fluoride removal from water. *Journal of Membrane Science*, 279, 529–538.

Jacks, G. & Battacharya, P. (2009) Drinking water from groundwater sources – a global perspective. In: Forare, J. (ed.) Drinking water: sources, sanitation and safeguarding. The Swedish Research Council Formas, Stockholm, Sweden. pp. 21–33.

Kabay, N. & Egawa, H. (1994) Chelating polymers for recovery of uranium from seawater. *Separation Science and Technology*, 28, 135–150.

Kanno, M. (1984) Present status of study on extraction of uranium from sea water. *Journal of Nuclear Science and Technology*, 21, 1–9.

Kantipuly, C., Katragadda, S., Chow, A. & Gesser, H.D. (1990) Chelating polymers and related supports for separation and preconcentration of trace metals. *Talanta*, 37, 491–497.

Kawai, T., Saito, K., Sugita, K., Katakai, A., Seko, N., Sugo, T., Kanno, J.I. & Kawakami, T. (2008) Comparison of amidoxime adsorbents prepared by cografting methacrylic acid and 2-hydroxyethyl methacrylate with acrylonitrile onto polyethylene. *Industrial & Engineering Chemistry Research*, 39, 2910–2915.

Khopkar, S.M. & Chalmers, R.A. (1970) *Solvent extraction of metals*. Van Nostrand Reinhold, London, UK.

Kilham, P. & Hecky, R.E. (1973) Fluoride: geochemical and ecological significance in East African waters and sediments. *Limnology and Oceanography*, 18, 932–945.

Koban, A. & Bernhard, G. (2004) Complexation of uranium(VI) with glycerol 1-phosphate. *Polyhedron*, 23, 1793–1797.

Kobuke, Y., Tabushi, I., Aoki, T., Kamaishi, T. & Hagiwara, I. (1988) Composite fiber adsorbent for rapid uptake of uranyl from seawater. *Industrial & Engineering Chemistry Research*, 27, 1461–1466.

Kryvoruchko, A.P., Yurlova, L.Y. & Yatsik, B.P. (2013) Influence of the structure of anionic surfactants on ultra and nanofiltration treatment of uranium-contaminated waters. *Radiochemistry*, 55, 123–128.

Kulkarni, P.S. (2003) Recovery of uranium(VI) from acidic wastes using tri-*n*-octylphosphine oxide and sodium carbonate based liquid membranes. *Chemical Engineering Journal*, 92, 209–214.

Kutahyali, C. & Eral, M. (2004) Selective adsorption of uranium from aqueous solutions using activated carbon prepared from charcoal by chemical activation. *Separation and Purification Technology*, 40, 109–114.

Lhassani, L., Rumeau, M., Benjelloun, D. & Pontie, M. (2001) Selective demineralization of water by nanofiltration: application to the defluorination of brackish water. *Water Research*, 35, 3260–3264.

Li, N.N., Fane, A.G., Ho, W.S.W. & Matsuura, T. (eds.) (2008) *Advanced membrane technology and applications*. John Wiley & Sons, Inc., Hoboken, NJ. p. 271.

Mahmoud, M.E. & Al Saadi, M.S.M. (2001) Selective solid phase extraction and preconcentration of iron(III) based on silica gel-chemically immobilized purpurogallin. *Analytica Chimica Acta*, 450, 239–244.

Meenakshi, S. & Viswanathan, N.J. (2007) Identification of selective ion-exchange resin for fluoride sorption. *Journal of Colloid and Interface Science*, 308, 438–450.

Min, B., Gill, A.L. & Gill, W.N. (1984) A note on fluoride removal by reverse osmosis. *Desalination*, 49, 89–93.

Mohsen, M.A. & Mohammed, F.A. (2013) Studies on uranium removal using different techniques. Overview. *Journal of Dispersion Science and Technology*, 34, 182–213.

Nalwa, H. (2001) *Advanced functional molecules and polymers*. Gordon and Breach Science Publishers, Singapore, Singapore. p. 323.

Nanyaro, J.T., Aswathanarayana, U. & Mungure, J.S. (1984) A geochemical model for the abnormal fluoride concentrations in waters in parts of northern Tanzania. *Journal of African Earth Sciences*, 2, 129–140.

Nath, S.K. & Dutta, R.K. (2010) Fluoride removal from water using crushed limestone. *Clean-Soil Air Water*, 38, 614–622.

OECD (2000) Uranium 1999: resources, production and demand. OECD Nuclear Energy Agency and the International Atomic Energy Agency, Paris, France.

OECD (2010) Uranium 2009: resources, production and demand. Publication 6891, OECD Nuclear Energy Agency, Paris, France.

Oliveria, E.E.M., Barbosa, C.C.R. & Afonso, J.C. (2012) Selectivity and structural integrity of a nanofiltration membrane for treatment of liquid waste containing uranium. *The Membrane Water Treatment*, 3, 231–242.

Oliveria, E.E., Barbosa, C.C.R. & Afonso, J.C. (2013) Stability of a nanofiltration membrane after contact with a low-level liquid radioactive waste. *Química Nova*, 36, 1434–1440.

Paugam, L., Diawara, C.K., Schlumpf, J.P., Jaouen, P. & Quéméneur, F. (2004) Transfer of monovalent anions and nitrates especially through nanofiltration membranes in brackish water conditions. *Separation and Purification Technology*, 40, 237–242.

Pontie, M., Diawara, C.K. & Rumeau, M. (2002) Streaming effect of single electrolyte mass transfer in nanofiltration: potential application for the selective defluorination of brackish drinking waters. *Desalination*, 151, 267–274.

Prabhakar, S., Balasubramaniyan, C., Hanra, M.S., Misra, B.M., Roy, S.B., Meghal, A.M. & Mukherjee, T.K. (1996) Performance evaluation of reverse osmosis (RO) and nanofiltration (NF) membranes for the decontamination of ammonium diuranate effluents. *Separation Science and Technology*, 31, 533–544.

Qadeer, R. & Hanif, J. (1994) Kinetics of uranium(VI) ions adsorption on activated-charcoal from aqueous solutions. *Radiochimica Acta*, 65, 259–266.

Raff, O. & Wilken, R.D. (1999) Removal of dissolved uranium by nanofiltration. *Desalination*, 122, 147–150.

Rana, D., Matsuura, T., Kassim, M.A. & Ismail, A.F. (2013) Radioactive decontamination of water by membrane processes – a review. *Desalination*, 321, 77–92.

Reimann, C. & Banks, D. (2004) Setting action levels for drinking water: are we protecting our health or our economy (or our backs!)? *Science of the Total Environment*, 332, 13–21.

Richards, L.A., Richards, B.S. & Schäfer, A.I. (2011) Renewable energy powered membrane technology: salt and inorganic contaminant removal by nanofiltration/reverse osmosis. *Journal of Membrane Science*, 369, 188–195.

Ritcey, G.M. & Ashbrook, A.W. (1984) *Solvent extraction: principles and applications to process metallurgy*. Part I. Elsevier Science Publishers, Amsterdam, The Netherlands.

Rivas, B.L., Pereira, E.D. & Villoslada, I.M. (2003) Water-soluble polymer–metal ion interactions. *Progress in Polymer Science*, 28, 173–179.

Schmitt, G.L. & Pietrzyk, D.J. (1985) Liquid chromatographic separation of inorganic anions on an alumina column. *Analytical Chemistry*, 57, 2247–2251.

Schwochau, K. (1984) Extraction of metals from sea water. *Topics in Current Chemistry*, 124, 91–133.

Sheppard, S.C., Sheppard, M.I., Gallerand, M.O. & Sanipelli, B. (2005) Derivation of ecotoxicity thresholds for uranium. *Journal of Environmental Radioactivity*, 79, 55–83.

Singh, H., Vijayalakshmi, R., Mishra, S.L. & Gupta, C.K. (2001) Studies on uranium extraction from phosphoric acid using di-nonyl phenyl phosphoric acid-based synergistic mixtures. *Hydrometallurgy*, 59, 69–76.

Solangi, I.B., Memon, S. & Bhanger, M.I. (2009) Removal of fluoride from aqueous environment by modified Amberlite resin. *Journal of Hazardous Materials*, 171, 815–819.

Tahaikt, M., El Habbani, R., Ait Haddou, A., Achary, I., Amor, Z., Taky, M., Alami, A., Boughriba, A., Hafsi, M. & Elmidaoui, A. (2007) Fluoride removal from groundwater by nanofiltration. *Desalination*, 212, 46–53.

Tahaikt, M., Ait Haddou, A., El Habbani, R., Amor, Z., Elhannouni, F., Taky, M., Kharif, M., Boughriba, A., Hafsi, M. & Elmidaoui, A. (2008) Comparison of the performance of three commercial membranes in fluoride removal by nanofiltration. Continuous operations. *Desalination*, 225, 209–219.

Tang, Z.J., Zhang, P. & Zuo, S.Q. (2003) Technology research of treatment for uranium waste water of low concentration. *Industry Water Waste*, 4, 9–12.

Teuler, A., Glucina, K. & Laine, J.M. (1999) Assessment of UF pretreatment prior RO membranes for seawater desalination. *Desalination*, 125, 89–96.

Tomar, V., Prasad, S. & Kumar, D. (2013) Adsorptive removal of fluoride from aqueous media using *Citrus limonum* (lemon) leaf. *Microchemical Journal*, 112, 97–103.

Turner, B.D., Binning, P. & Stipp, S.L.S. (2005) Fluoride removal by calcite: evidence for fluorite precipitation and surface adsorption. *Environmental Science & Technology*, 39, 9561–9568.

UN (2013) Press release embargoed until 13 June 2013, 11:00 A.M., EDT.

UNICEF (1999) State of the art report on the extent of fluoride in drinking water and the resulting endemicity in India. Report by Fluorosis Research & Rural Development Foundation of UNICEF, New Delhi, India.

Van der Bruggen, B., Koninckx, A. & Vandecasteele, C. (2004) Separation of monovalent and divalent ions from aqueous solution by electrodialysis and nanofiltration. *Water Research*, 38, 1347–1353.

Villalobos-Rodríguez, R., Montero-Cabrera, M.E., Esparza-Ponce, H.E., Herrera-Peraza, E.F. & Ballinas-Casarrubias, M.L. (2012) Uranium removal from water using cellulose triacetate membranes added with activated carbon. *Applied Radiation and Isotopes*, 70, 872–881.

WHO (2006) *Fluoride in drinking-water*. World Health Organisation, Geneva, Switzerland. Available from: www.who.int/water_sanitation_health/publications/fluoride_drinking_water_full.pdf [accessed January 2014].

Yaman, M. & Gucer, S. (1995) Determination of Cd and Pb in vegetables after activated carbon enrichment by atomic absorption spectrometry. *Analyst*, 120, 101–107.

Yamini, Y., Chaloosi, M. & Ebrahimzadeh, H. (2002) Solid phase extraction and graphite furnace atomic adsorption spectrometric determination of ultra trace amount of bismuth in water samples. *Talanta*, 56, 797–803.

Yurlova, L.Y. & Kryvoruchko, A.P. (2010) Purification of uranium containing waters by the ultra and nanofiltration using modified montmorillonite. *Journal of Water Chemistry and Technology*, 32, 358–364.

CHAPTER 4

The use of reverse osmosis (RO) for removal of arsenic, fluoride and uranium from drinking water

Priyanka Mondal, Anh Thi Kim Tran & Bart Van der Bruggen

4.1 REVERSE OSMOSIS: BACKGROUND AND TRANSPORT MECHANISM

RO was used for removal of salts from water and was first patented on 1931 by Horvath. In this process a fluid is passed through a semi permeable membrane from a higher solute concentration compartment to lower solute concentration compartment by applying pressure. In 1959, a remarkable improvement of this process was established by Reid and Breton as they achieved 98% of salt rejection by applying a cellulose acetate membrane. However, due to the large thickness of the membrane (5–20 μm) the resultant flux was not high. The use of anisotropic cellulose acetate membranes (Loeb and Sourirajan, 1963) increased the applicability of RO in the desalination industry due to its improved performance. Improved composite membranes (made by interfacial polymerization) (Cadotte, 1977) also made a large breakthrough for economical application of RO membrane in drinking water industry. The RO membranes are also able to exclude lower molar mass species (salt ions, organics etc.) due to their non-porous structure. Some membranes have a performance between ultrafiltration and RO due to their porous structure; these are known as "loose RO membranes". However, the pore size of those membranes is very small (in the order of 10 Å or less) (Baker, 2004). Nowadays, several improvements are being implemented on conventional RO membranes (detailed discussion in Section 4.2) and also in RO plants, to make the system more sustainable. Although at present, RO is used mainly in desalination, further improvements are still required for broader applications of this technique.

RO membranes consist of two layers, (i) a thin dense top polymer layer and (ii) a porous sublayer, which gives support to the top layer and increases the mechanical stability of the membrane. These membranes have been effectively used for water desalination with a very high rejection (sometimes over 99%) of the low molecular mass compounds (inorganic salts or small organic molecules) (Velizarov *et al.*, 2004). However, there are some drawbacks for RO treatment, which include (i) lack of dissolved minerals in the treated water, (ii) low rejection of neutral molecules and (iii) high energy consumption (Mondal *et al.*, 2013).

The transport of salts and water through RO membrane is controlled by the solution-diffusion mechanism and the flux of water (J) is calculated as:

$$J = A(\Delta p - \Delta\pi) \tag{4.1}$$

where A is a constant, Δp is difference across the membrane and $\Delta\pi$ is osmotic pressure difference across the membrane. Equation (4.1) explains three phenomenon; (i) when $\Delta\pi > \Delta p$, osmosis takes place, which implies that water flows from the dilute side to the concentrated side, (ii) $\Delta p = \Delta\pi$, in this case flux of water is not possible, and (iii) $\Delta\pi < \Delta p$, this condition implies that water flows from the higher concentration compartment to the lower concentration compartment, through a semi permeable membrane. Thus, the water flux is proportional to the applied pressure. However, the salt flux is independent of pressure. These three conditions are schematically represented in Figure 4.1. The overall rejection (R) of the salts is given by:

$$R = [1 - (C_{\text{permeate}}/C_{\text{feed}})] \tag{4.2}$$

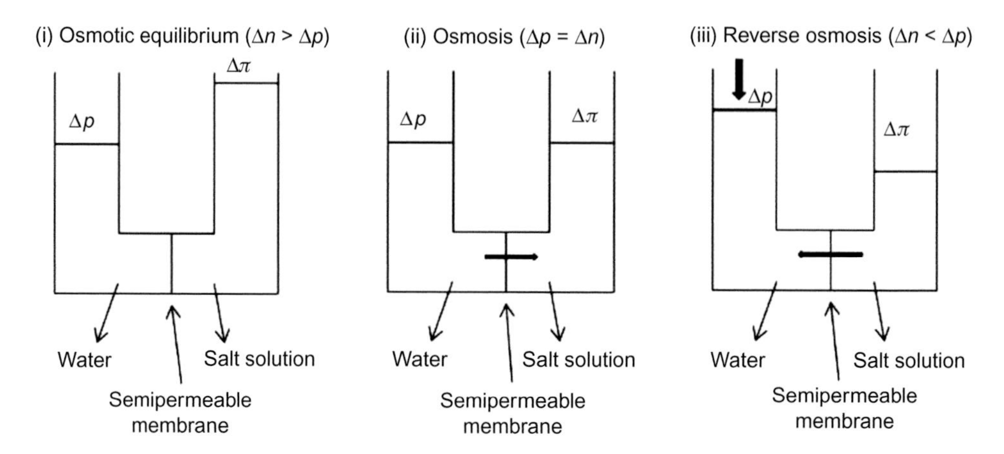

Figure 4.1. A schematic representation of (i) osmotic equilibrium, (ii) osmosis and (iii) reverse osmosis.

where C_{permeate} is the concentrations of salts at the permeate side and C_{feed} is the concentration of salts at the feed side of the membrane. However, there are several physical parameters that affect the salt rejection such as, concentration of salts in the feed chamber, feed water temperature, operating pressure and pH of the feed solution.

4.2 TYPES OF RO MEMBRANES

RO membranes are divided into three categories, i.e., cellulose acetate based membranes, polyamide based membranes and composite membranes. All of these three kinds of membranes are discussed below.

4.2.1 *Cellulose acetate membranes*

Cellulose acetate was the first material used to make RO membranes. This kind of material was first used by Loeb and Sourirajan in 1963. Nowadays the use of these membranes is limited due to their lower performance than composite membranes. Cellulose acetate membranes are inexpensive, very easy to prepare, resistant against oxidants and mechanically tough in nature. The membrane has an asymmetric or an anisotropic structure and consists of a thin active layer on a coarse supportive layer. However, this kind of membrane is very sensitive towards the pH and temperature of the feed water. Thus, it is better to maintain the feed water pH between 4 and 6 because the membranes are slowly hydrolyzed with time and above 35°C, the properties of the membrane change (Vos *et al.*, 1966). Moreover, these types of membranes are very susceptible to biological attack.

4.2.2 *Aromatic polyamide membranes*

Aromatic polyamide membranes are also known as non-cellulosic polyamide membranes. Several polymer materials were proposed to overcome the problems related to cellulose acetate membranes but only the use of polyamide proved successful (Wang *et al.*, 2011). Due to the lower rejection and relatively lower flux, aliphatic polyamide membranes are ruled out by aromatic polyamide membranes, which have a significantly higher rejection and are successfully produced commercially (Endoh *et al.*, 1977; McKinney and Rhodes, 1971; Strathmann, 1990). These membranes are asymmetric in nature and have very high salt rejections.

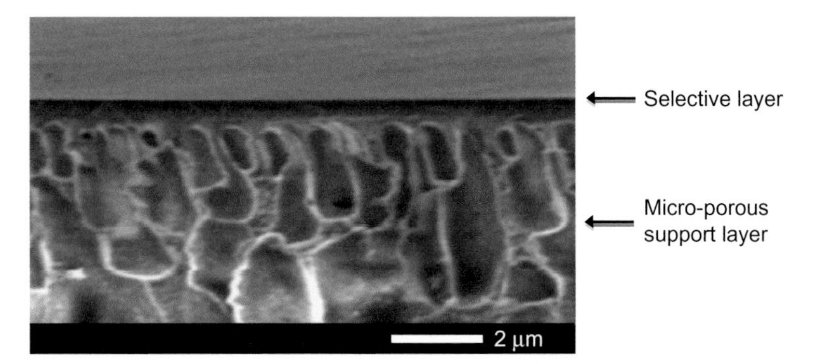

Figure 4.2. Scanning electron microscopic image of composite membrane (adapted from Perera *et al.*, 2014).

4.2.3 *Composite membranes*

Composite membranes are made of two or more polymeric materials. These types of membranes are prepared by coating microporous support layers with two or more active dense very thin ($\leq$0.1 μm) polymer layers (Fig. 4.2). The first composite membrane for RO was invented and patented by Cadotte in 1981. These membranes have a very high flux, high salt rejection, and are resistant over a broad range of pH (4–11) and can be operated above 35°C. However, the chlorine resistance of these membranes is very poor, which decreases the selectivity of membranes and makes them very difficult to use with chlorine contaminated (few $\mu g\, L^{-1}$) feed water (Baker, 2004).

4.3 MEMBRANE MODULES AND THEIR APPLICATION

A filtration device consists of a filtrate outlet structure and a membrane unit in which a specific membrane surface area is housed, the membrane module. There are four types of membrane modules that are used for RO: plate and frame, tubular, spiral wound and hollow fiber. A brief description about these membrane modules and their functions is given in the following sub sections. Several advantages and disadvantages of all these membrane modules are also discussed in Table 4.1.

4.3.1 *Plate and frame modules*

This type of module consists of a membrane, which is placed between the feed spacer and product spacer and then the system is attached to both sides of a rigid plane or end plate (Fig. 4.3). The plates are made of different materials, e.g., porous fiberglass, solid plastic with grooved channels on the surface, reinforced porous paper, etc. The membrane unit is placed in a pressurized vessel. Feed water is forced to pass across the surface of the membrane for filtration. Permeates are collected in product manifold after passing through the membrane and brine solution is collected from porous media. However, this membrane module is not used in large scale applications because of the complexity and cost of operation.

4.3.2 *Tubular modules*

In tubular modules, a number of tubes are arranged in series (Fig. 4.4). The tubes are made with porous paper or fiberglass support and the membranes are installed inside the tubes. The water is circulated through the tube under pressure and the filtered water is collected from collecting tube installed inside the system. Although the system can operate in extreme turbid feed water

Table 4.1. Advantages and disadvantages of several membrane modules.

Membrane modules	Advantages	Disadvantages
Plate and frame	(i) Applicable for high viscous liquid and can be operated at high pressure (ii) Easy to replace and clean the membrane	(i) Membrane working area is very small
Tubular	(i) Good resistivity with fouling (ii) Comparatively easy to clean	(i) Packing density is very low (ii) Energy consumption is highest with respect to other modules
Spiral wound	(i) Packing density is high (ii) Cost is relatively low (iii) Easy to adjust with hydrodynamics by changing the spacer thickness and can avoid the risk of fouling	(i) Difficult to clean (ii) Pressure drop is large
Hollow fiber	(i) Very high packing density (surface area to volume ratio) (ii) Energy consumption relatively low	(i) Very high capital cost (ii) Fouling resistivity is poor (iii) Difficult to clean and change the membrane

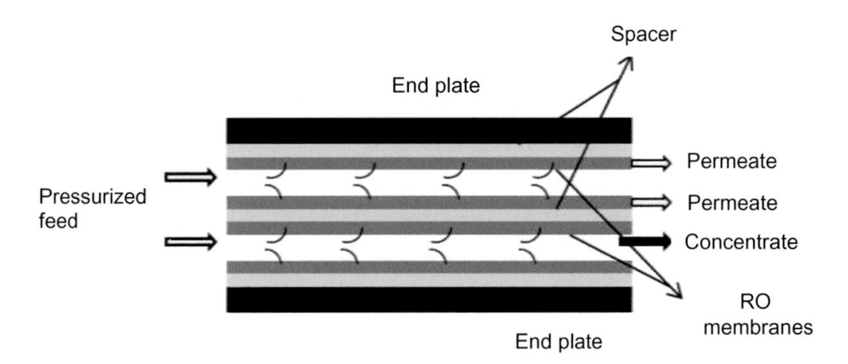

Figure 4.3. Schematic diagram for plate and frame module used in RO system.

and is easy to clean (mechanically or hydraulically), it is not the best option for RO due to its high capital cost.

4.3.3 *Spiral wound modules*

Spiral wound membrane modules are well-known in water desalination and are better than plate frame and tubular modules due to their high water flux, lower salt permeability and lower operational cost. Several membrane elements are connected with each other in a spiral mode and wrapped around a centrally installed permeate tube (Fig. 4.5). The whole setup is placed inside a pressurized tubular vessel where the feed water passes through the membranes axially down the module.

4.3.4 *Hollow fiber module*

Hollow fiber membrane modules consist of large number of hollow fibers that are asymmetric in nature (Fig. 4.6). Usually hollow fiber membrane modules are formed in two geometries, (i) a small diameter and thick wall fiber bundle (internal diameter 50 μm and outer diameter 100 to

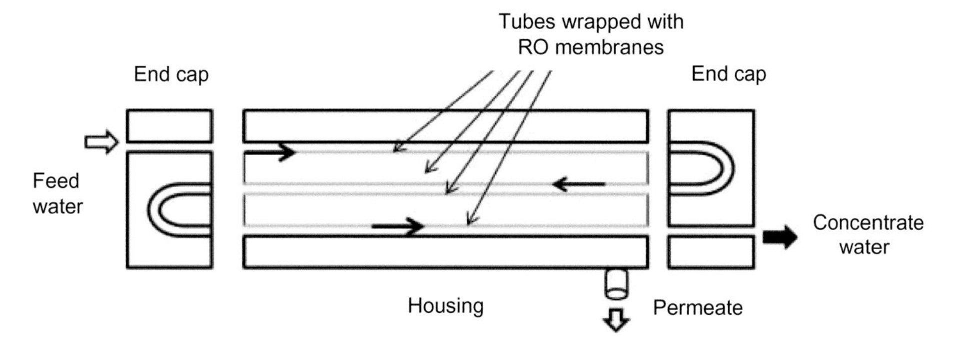

Figure 4.4. Schematic representation of tubular RO membrane module.

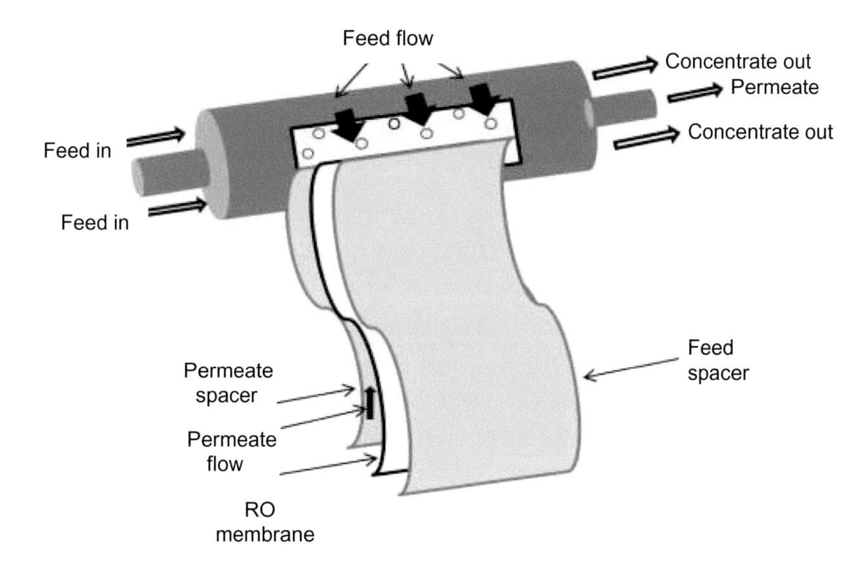

Figure 4.5. Schematic representation of spiral wound RO membrane module.

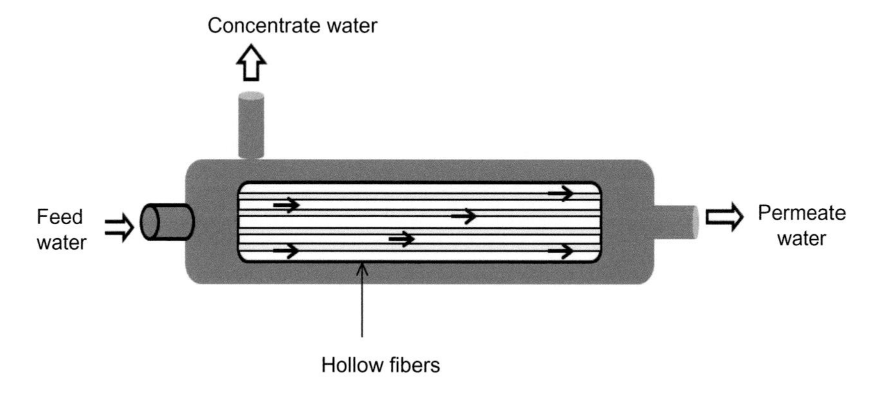

Figure 4.6. Schematic representation of hollow fiber RO membrane module.

200 μm) contained in a vessel where the feed water passes through the fiber wall and permeates are collected through open fiber end, and (ii) feed water passes through the bore of the fiber, which is open at both ends (known as bore side feed type hollow fiber membrane) (Baker, 2004). The pressurized feed water flows radially over the fiber bundle and the permeate is collected outside of the fiber bundle into the hollow fiber bore. This module serves a very large surface area for polluted water purification. However, this module is very susceptible to fouling due to its compact structure.

4.4 REVERSE OSMOSIS FOR ARSENIC, FLUORIDE AND URANIUM REMOVAL FROM WATER

RO is a well-established technology that has been used successfully to remove As, F and U from contaminated water for many years. As and F generally occur as an anionic species in experimental pH conditions. Similarly, zeta potential values of the applied RO membranes are also negative within the studied pH range. Therefore, better rejection was observed due to charge exclusion (Donna exclusion). However, not only charge exclusion, size exclusion (especially for U) and diffusion coefficients play an important role for the removal of these contaminants. A brief summary of the removal efficiency and experimental details of these contaminants by using RO membranes are elaborated in the following sections.

4.4.1 *RO for removal of arsenic from water*

Several studies have been performed on removal of As from water by RO. These are summarized in Table 4.2.

Yoon *et al.* (2009) used a bench-scale cross-flow flat-sheet filtration system with an LFC-1 (Hydranautics), polyamide TFC RO membrane to remove As(III) and As(V) from model and natural waters. They found a high removal of As(V) and the rejection enhanced (90–99%) with increasing pH (from 4–10) due to the higher degree of dissociation of fixed ionizable functional groups on the membrane. Moreover, they also found that the rejection of As(III) by the same RO membranes was constant (92–96%) and the change in pH did not affect the rejection due to dominance of steric exclusion over charge repulsion. Teychene *et al.* (2013) observed that the SW membrane has better removal efficiency for As(III) than the BW membrane. They concluded that the rejection of the metalloid depends on the applied transmembrane pressure, characteristics of the membrane and the pH of the solution. A similar conclusion was also suggested by Akin *et al.* (2011). Low pressure polyamide RO membranes were used by Deowan *et al.* (2008). They found that the rejection of As(V) (over 98%) was higher than that of As(III) (60–80%) from As spiked local tap water. The rejection of As(V) by the LE membrane was high and reached below the MCL up to a feed concentration of 2000 $\mu g\,L^{-1}$. For the XLE membrane, the maximum feed concentration was 800 $\mu g\,L^{-1}$ of As(V) above which the permeate concentration exceeded the MCL. On the other hand, the rejection of As(III) below the MCL was possible if the initial feed concentration was up to 50 $\mu g\,L^{-1}$. In another study, a small scale RO unit for the removal of both As(V) and As(III) using spiked local tap water was studied by Geucke *et al.* (2009). In conventional RO systems, the energy from the rejected water is usually wasted, but the RO pump (Fig. 4.7) used in this study could reuse the energy of the rejected water and thus, reduced the expenses by 50–90%. However, they also observed that in case of As(III), the permeate concentrations reached below the MCL only when the feed concentration was below 350 $\mu g\,L^{-1}$ for As(III), whereas for As(V), the limit was up to 2400 $\mu g\,L^{-1}$. Thus, these studies suggest a better removal efficiency of As(V) than As(III) in RO systems mainly due to electrostatic repulsion between RO membranes and As(V) species. The rejection of As(III) is low because of its neutral characteristics.

Floch *et al.* (2004) studied the removal of As in a pilot plant using a polyvinylidene fluoride RO membrane (ZW-1000; ZENON Environmental Inc. Canada), after several pretreatment

Table 4.2. Arsenic removal by RO.

Type of RO membrane and manufacturer	Origin of water	Rejection [%]		Operating pH	Flux [$m^3\ m^{-2}\ h^{-1}$]
		As(V)	As(III)		
SW30HR (DOW, FilmTec)	Synthetic brackish water (Teychene *et al.*, 2013)	–	>99	7.6	18.72 (at 2.4 MPa)
SCW5 (Hydranautics)		–	>99	7.6	27.84 (at 2.4 MPa)
BW30LE (DOW, FilmTec)		–	99	9.6	85.6 (at 4.0 MPa)
ESPAB (Hydranautics)		–	99	9.6	102.8 (at 4.0 MPa)
ESPA2 (Hydranautics)		–	99	9.6	118.4 (at 4.0 MPa)
XLE-2521 (DOW, FilmTec)	As spiked local tap water (Geuke *et al.*, 2009)	96.1–>99	73.2–97	7.2	28.3 (at 0.45 MPa)
TW30-2521 (DOW, FilmTec)					26.7 (at 0.96 MPa)
SW30-2521 (DOW, FilmTec)					25.8 (at 1.52 MPa)
ZW 1000 (Zenon)	Well water (Floch *et al.*, 2004)	Total As rejection 97–>99		–	–
SWHR (DOW, FilmTec)	Spiked water and Natural groundwater sample (Akin *et al.*, 2011)	96	83	2.4–10.4	11.60 and 7.07 (at 1.0–3.5 MPa) for As(V) and As(III), respectively
BW-30 (DOW, FilmTec)		78	68		0.70 and 0.28 (at 1.0–3.5 MPa) for As(V) and As(III), respectively
LE (DOW water)	Arsenic spiked local tap water (Deowan *et al.*, 2008)	>95	60–80	5–9	40–60
XLE (DOW water)		>95	60–80		40–60
BW30 (DOW, FilmTec)	Australian groundwater (Richards *et al.*, 2009)	Total As removal 79		3–11	22.5
ESPA4		Total As removal 75			

processes such as oxidation with potassium permanganate and coagulation with ferrous sulfate. The pretreatment units were separated into three parts prior to the application of membrane filtration, and the As concentration in the permeate was below the MCL. However, the As contaminated pretreatment waste may be a greater concern for the environment.

A photovoltaic-powered RO desalination system combined with a two-staged membrane system (UF and RO) was studied by Richards *et al.* (2009). They used two RO membranes for removal of As from Australian groundwater and obtained a high rejection of total As. The flux during the operation decreased from 24.2 to 22.5 $L\ m^{-2}\ h^{-1}$ due to precipitation on the surface of the membrane. Their study suggested that a proper understanding of the dominant aqueous species, along with the proper choice of membranes and the use of renewable energy sources can resolve the problem of water scarcity in many parts of the world.

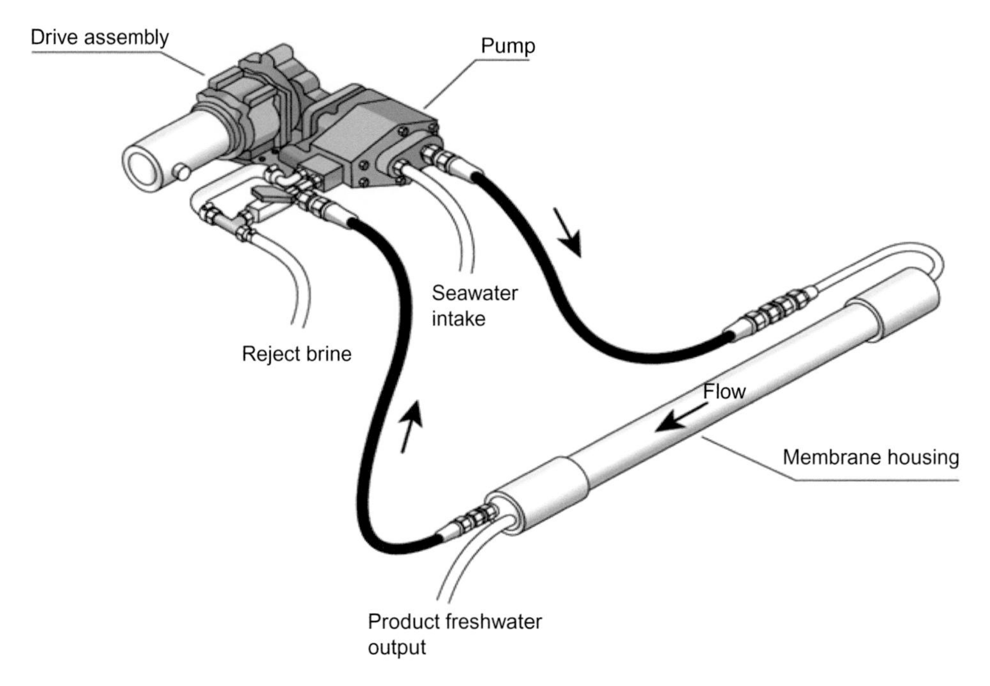

Figure 4.7. RO pump used by Geuke *et al.* (2009) (adapted from Bhattacharya *et al.*, 2009) (Katadyn Power Survivor, 160E, 2009).

All the studies related to removal of As from aqueous solution are summarized in Table 4.2. These studies prove that although RO can remove As(V) below the MCL, limitations are related to the occurrence of As(III) at nearly neutral pH.

4.4.2 *Fluoride removal from water by RO*

Reported studies on F^- removal from aqueous solution by RO are summarized in Table 4.3.

The pH is an important parameter for the removal of F^- by RO. Richards *et al.* (2010) studied the dependence of F^- retention on pH with two RO membranes, BW30 (RO, Dow, FilmTec), UTC-80A (RO, Toray). The results demonstrate that at acidic pH, the retention of F^- was low due to the dominance of HF, and is much different among the 6 tested membranes. On the contrary, at pH > 7, with Cl^-, HCO_3^- and Na^+ as co-existing ions, the rejection was more than 90% for both the RO membranes.

In a different study, Dolar *et al.* (2011) studied the F^- removal efficiency with three RO membranes to treat fertilizer wastewater. With RO membranes LFC-1 (assumed pore size 0.75 nm), XLE (0.62 nm) and ULP (0.73 nm), the efficiency was above 96%.

Apart from studies on a laboratory scale, F^- removal was also successfully applied on a pilot scale. Sehn (2008) used a pilot plant with a RO membrane (capacity of 6000 $m^3\ d^{-1}$) and demonstrated that RO can remove F^- from groundwater below 0.03 $mg\ L^{-1}$ (the rejection was 98.4%). The rejection remained unchanged after 3 years of operation. Similarly, Diawara *et al.* (2011) investigated F^- removal from brackish ground water for the 2000 inhabitants of the rural community of Ndiaffate (Kaolack, Senegal). The pilot plant low pressure reverse osmosis (LPRO) from Dow Company was designed with a feed flow rate of 900 $L\ h^{-1}$ and a conversion rate of 66%. With the LPRO, the rejection was higher, i.e., between 97 and 98.9%. Thus, LPRO appeared to be an effective method to treat the water to meet drinking water standards.

Table 4.3. Fluoride removal by RO.

Membrane manufacturer	Water origin	Rejection [%]	Flux [$L\,m^{-2}\,h^{-1}$]
XLE (Dow/Filmtec) (Sehn, 2008)	Groundwater	98.4	n/a (at 0.6–1.1 MPa)
BW30 (RO, Dow, FilmTec) (Richard *et al.*, 2010)	Water spiked with NaF, NaCl and $NaHCO_3$	90–95 (pH > 7)	12.1 (at 0.5 MPa)
UTC-80A (RO, Toray) (Richard *et al.*, 2010)		90–95 (pH > 7)	15.4 (at 0.5 MPa)
LFC-1 (RO, Hydranautics) (Dolar *et al.*, 2011)	Real wastewater from fertilizer factory	96.8	25.57 (at 2.5 MPa)
XLE (RO, Dow, FilmTec) (Dolar *et al.*, 2011)		96	26.96 (at 2.5 MPa)
ULP (RO, Koch Membrane) (Dolar *et al.*, 2011)		96.6	30.51 (at 2.5 MPa)
LPRO (Dow Chemical) (Diawara *et al.*, 2011)	Brackish groundwater	97–98.9	79 ($Y = 66\%$) (Y is the conversion rate)

4.4.3 *RO for uranium removal from water*

Several water treatment technologies have been used for removal of U below the MCL including activated carbon (Coleman *et al.*, 2003), coagulation-flocculation (Gafvert *et al.*, 2002), ion exchange (Barton *et al.*, 2004), adsorption (Shuibo *et al.*, 2009) and ultrafiltration with precomplexation (Kryvoruchko *et al.*, 2004). However, very few studies focused on removal of U from contaminated water by using RO.

Removal of U from brackish groundwater of Australia using direct solar powered ultrafiltration-nanofiltration/reverse osmosis membrane system was studied by Rossiter *et al.* (2010) (Fig. 4.8). The feed pressure and feed flow were 0.9 MPa and 400 $L\,h^{-1}$, respectively. They found that U was strongly adsorbed to membranes at pH 4–7 but retained by the membranes over the pH range of 3–11. The retention of U including other divalent cations was >99% for batch experiments and 95–98% for continuous experiments. However, during the solar energy experiments, they observed a decreased retention of U (above WHO limit), which was because of precipitation of U on the membrane. In another study by Montaña *et al.* (2013) more than 90% U removal from contaminated river water was obtained by using an RO pilot plant. They also found that the heavy molecular weight U complexes were rejected by dense RO membranes *via* molecular filtration. RO can remove 99% U with higher water desalination and therefore is a better option than other technologies (Gamal Khedr, 2013).

4.5 EXPERIMENTAL STUDY FOR REMOVAL OF ARSENIC AND FLUORIDE FROM WATER BY RO AND LOOSE RO MEMBRANES

4.5.1 *Materials and methods*

4.5.1.1 *Membranes*

RO experiments with low pressure and high pressure membranes were performed by using four types of commercial membranes supplied by Dow-FilmTec with flat sheet configuration having an active surface area of 59 cm^2. For high pressure RO experiments SWHR and BW-30 flat sheet membranes were used. Similarly, NF90 and NF 270, two dense nanofiltration membranes were used for low pressure RO experiments.

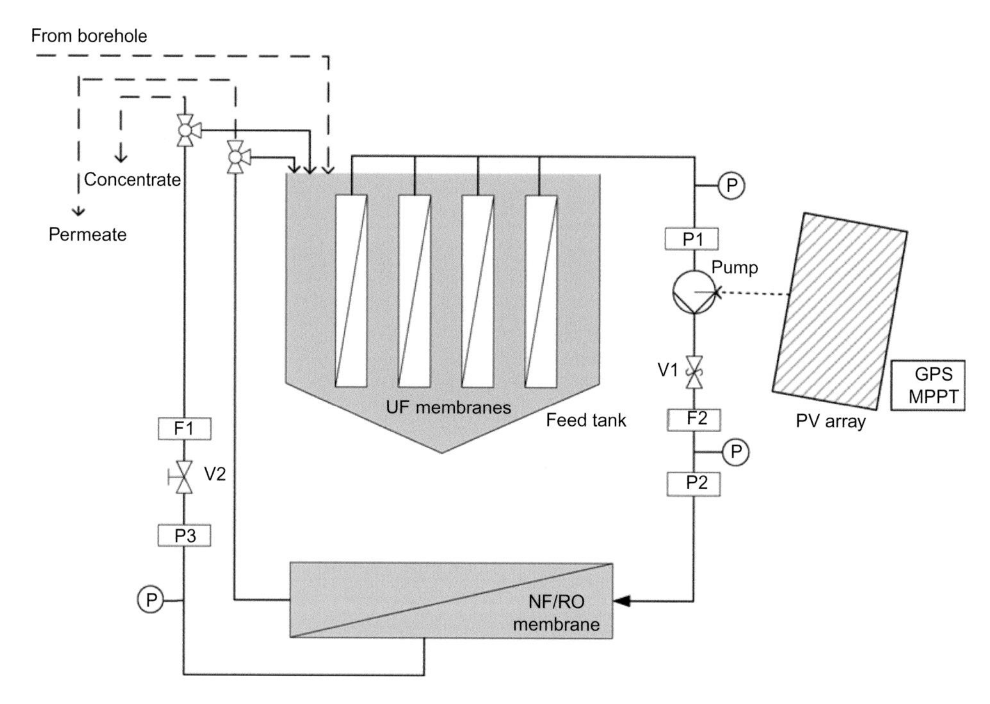

Figure 4.8. Solar powered membrane filtration system for uranium removal (adapted from Rossiter *et al.*, 2010) where P = pressure gauges; P1–3 = pressure transducers; F1–2 = flow sensors; V1 = pressure relief valve; V2 = pressure control valve; GPS = solar tracker guided by global positioning system and MPPT = maximum power point tracker.

4.5.1.2 *Reagents and chemicals*

The solution of pentavalent arsenic and monovalent fluorine was prepared by dissolving analytical grade $Na_2HAsO_4{\cdot}7H_2O$ (RPL, Belgium) and NaF (Riedel-deHaën) in distilled water. Arsenic and F^- standard solutions, with concentration ranging 50–300 $\mu g\,L^{-1}$ and 2.5–15 $mg\,L^{-1}$ respectively were prepared by dilution, immediately before starting the experiment. All stock solutions were prepared using deionized (DI) water (18.2 $m\Omega\,cm^{-1}$) from a Mili-Q water system. The pH of the solution (7 ± 0.2) was adjusted by either HNO_3 or NaOH.

4.5.1.3 *Groundwater characteristics*

The synthetic groundwater was prepared by following the northeastern India groundwater composition mentioned in Singh *et al.* (2008). In brief, synthetic groundwater was prepared by adding $NaNO_3$ (Riedel-deHaën), Na_2SO_4 (Acros), $MgCO_3$ (Sigma-Alrrich) and $CaSO_4{\cdot}2H_2O$ (Riedel-deHaën) into deionized water. Arsenic and F^- solutions were spiked from a 1000 $mg\,L^{-1}$ stock solution. The initial concentrations of the ions in the synthetic groundwater were as following: Ca^{2+}: 180 $mg\,L^{-1}$, Mg^{2+}: 62 $mg\,L^{-1}$, Na^+: 140 $mg\,L^{-1}$, SO_4^{2-}: 385 $mg\,L^{-1}$, As(V): 0.2 $mg\,L^{-1}$, F^-: 10 $mg\,L^{-1}$ and NO_3^-: 220 $mg\,L^{-1}$.

4.5.1.4 *Membrane performance*

All the experiments for the removal of As(V) and F^- were carried out for 6 hours and each time a new membrane was used in each experiment. The rejections of all the ions were calculated according to the following equation:

$$\text{Rejection [\%]} = (1 - (C_{permeate}/C_{feed})) \times 100 \tag{4.3}$$

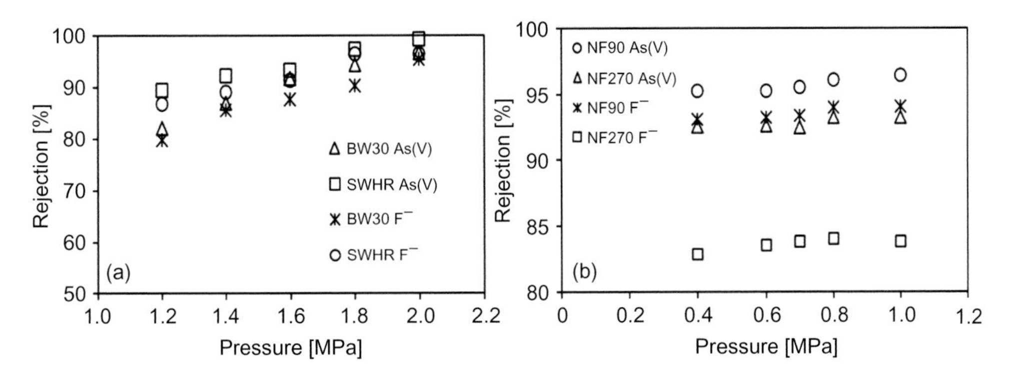

Figure 4.9. The rejection of As(V) and F^- as a function of pressure for (a) RO and (b) NF. Operational conditions: As(V) = 150 $\mu g\,L^{-1}$, $F^- = 7.5\,mg\,L^{-1}$, pH = 7 ± 0.2, Temperature = 20°C).

where $C_{permeate}$ and C_{feed} are the concentration of the ions [$mg\,L^{-1}$] in permeate and feed, respectively.

4.5.1.5 *Analytical methods*

Samples of the feed and permeate from the membrane system were taken at 30 min intervals and made ready for analysis. The concentrations of cations (As^{+5}, Ca^{2+}, Mg^{2+} and Na^+) were determined by ICP-MS (Thermo Electron Corporation X series ICP-MS) and anions (F^-, NO_3^-, SO_4^{2-}) were determined by IC-DIONEX (ICS 2000). The pH was fixed at 7 ± 2 for all the experiments and measured by an Orion pH meter (USA).

4.5.2 *Results and discussion*

4.5.2.1 *Effect of operating pressure*

Figure 4.9 shows that the removal of both As(V) and F^- in RO increases with increasing operating pressure. With RO, the rejection of As(V) increased from 82.1 to 96.7% and from 89.4 to 99.4% when using the BW30 and SWHR membranes, respectively, whereas for F^- the rejection increased from 79.7 to 95.4% and from 86.8 to 96.5% for the BW30 and SWHR membranes, respectively, by increasing the pressure from 1.2 to 2.0 MPa. For all operating pressure conditions, SWHR membranes provided higher As(V) and F^- rejections (with a large difference) and thus, in comparison with BW30 membranes, SWHR membranes were shown to have higher rejection efficiencies for both As(V) and F^- (Fig. 4.9). Similar results were obtained by Akin *et al.* (2011) and Gholami *et al.* (2006) for As(V). However, the F^- rejection obtained with the BW membrane was lower than reported in the literature (98.4%) (Sehn, 2008). Since As(V) was present in the solution as divalent $HAsO_4^{2-}$, which is strongly repulsed by the membranes and increased the ionic strength, the divalent $HAsO_4^{2-}$ forces F^- to pass through the membranes to maintain the electroneutrality and ultimately decreases the rejection (Choi *et al.*, 2001).

Similarly, the rejection of both As(V) and F^- by NF90 was higher than that of NF270 over the investigated pressure range (Fig. 4.9). The removal of As and F with the NF90 membrane was higher than 96% and 94%, while for the NF270 membrane the rejection was higher (than 93% and 84%, respectively). The lower molecular weight cutoff of the loose NF270 membrane (300 Da) in comparison with the more dense NF90 membrane (200 Da) (Hoinkis *et al.*, 2011) causes a lower rejection of NF270. The increase in operating pressure did not improve the As rejection significantly, which was similar to the findings of Saitua *et al.* (2011) and Figoli *et al.* (2010). The rejection of F^- decreases with increasing pore size of the membrane (Fig. 4.9) from NF 90 (~93%) (pore size 0.68 nm) (Nghiem and Hawkes, 2001) to NF 270 (~83%) (pore size ~0.84 nm) (Nghiem and Hawkes, 2007) because the looser membrane (NF 270) has a lower rejection than

Table 4.4. Ions and their respective diffusion coefficient and hydrated radii.

Ions	Diffusion coefficient [10^{-5} cm^2 s^{-1}]	Hydrated radii [Å]
Na^+	1.33 (Linde and Jönson, 1995)	3.58 (Nightingale Jr, 1959)
Mg^{2+}	0.71 (Linde and Jönson, 1995)	4.28 (Nightingale Jr, 1959)
Ca^{2+}	0.92 (Vrijenhoek and Waypa, 2000)	4.12 (Nightingale Jr, 1959)
NO_3^-	1.902 (Wang *et al.*, 2005)	3.35 (Nightingale Jr, 1959)
F^-	1.45 (Atkins, 1998)	3.52 (Nightingale Jr, 1959)
$HAsO_4^{2-}$	0.323 (Vrijenhoek and Waypa, 2000)	>2.0–2.2 (Robinson and Stokes, 1965)
SO_4^{2-}	1.065 (Vrijenhoek and Waypa, 2000)	3.79 (Nightingale Jr, 1959)

the tight membrane (NF 90). Dolar *et al.* (2011) found asimilar rejection of F^- by the tight NF90 membrane (98.9%) and the loose TFC membrane (75.4%). The removal of divalent $H_2AsO_4^{2-}$ was always higher than that of monovalent F^- for balancing the electroneutrality. This is also related to the low diffusion coefficient of $H_2AsO_4^{2-}$ compared to F^- (Table 4.4). However, the As and F^- concentration in the permeate of both membranes was always below the MCL.

4.5.2.2 *Effect of feed water concentration*

The rejection of As(V) and F^- as a function of feed water concentration is shown in Figure 4.10. The experiments were performed at 0.8 and 1.8 MPa for low pressure RO and high pressure RO membranes, respectively, and at a temperature of 20°C. There was no significant change observed for As(V) rejection with change of the concentrations (96.1–96.4% for BW30 and 97.8–98.6% for SWHR). Because of the increase of the feed water concentration, the concentration in permeate also increased. Thus, the permeate concentration of As remained nearly constant. The rejection of As(V) by the SWHR membrane was always higher than that of the BW30 membrane (Fig. 4.10a). This result is in agreement with what has been reported by Gholami *et al.* (2006) and Akin *et al.* (2011). However, a slight decrease of rejection with increasing concentrations was observed for F^- with both membranes (88.6–87.1% for BW30 and 92.1–91.2% for SWHR) due to the high ionic concentration (Fig. 4.10b). The concentration of F^- was below the detection limit (BDL) with an initial concentration of 2.5 mg L^{-1} and 5 mg L^{-1}, and remained below the MCL for all the other concentrations. Dolar *et al.* (2011) also reported similar results for F^- removal using the BW membrane.

Figures 4.10c and 4.10d show the effect of As(V) and F feed concentrations for both low pressure RO membranes at an operating pressure of 0.8 MPa and a temperature of 20°C. The NF90 membrane (over 96%) showed a higher As rejection than the NF270 membrane (over 92%), in the investigated As feed concentration range. The rejection of As(V) by both membranes decreased slightly (NF90 99–96% and NF270 95–92%) with increasing concentration from 50 μg L^{-1} to 300 μg L^{-1} (Fig. 4.10c). This is in agreement with Akbari *et al.* (2011), who found a decrease of the As rejection with the NF90 membrane from 98.3 to 96.6%, by increasing the feed concentration. Furthermore, the rejection of F^- by the NF90 membrane (over 95%) was higher than for the NF 270 membrane (over 84%), and the rejection was decreased with increasing feed concentration (95–92% NF90 and 84–84.2% NF270) (Fig. 4.10d).

4.5.2.3 *Removal of As and F from synthetic groundwater*

Figure 4.11 shows the total rejection of several ions including As(V) and F^- by two low pressure RO membranes and two high pressure RO membranes. The rejection for As(V) ranged between 96.1 and 98.6% and for F^- between 84.4 and 97.2%. For all four membranes, the rejection sequence of cations was $Ca^{2+} \geq Mg^{2+} > Na^+$ and for the anions $SO_4^{2-} > HAsO_4^{2-} > F^- > NO_3^-$. For cations, the higher rejection of divalent cations (Ca^{2+} and Mg^{2+}) compared to monovalent

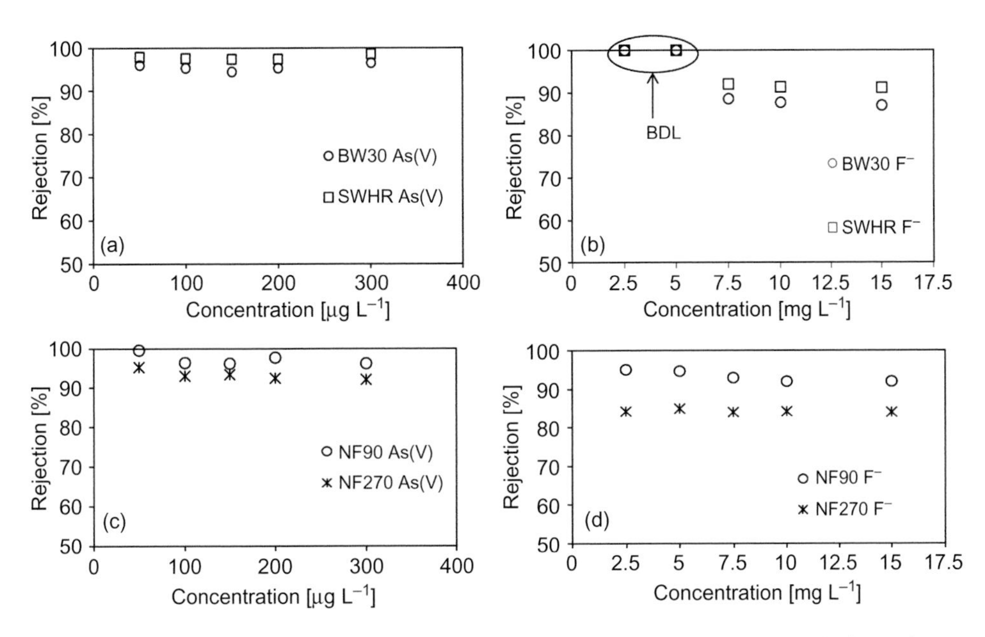

Figure 4.10. Rejection of (a) As(V) and (b) F^- of RO and rejection of (c) As(V) and (d) F^- for NF as a function of concentration. (Operational conditions: pressure 1.8 MPa, pH = 7 ± 0.2, temperature = 20°C).

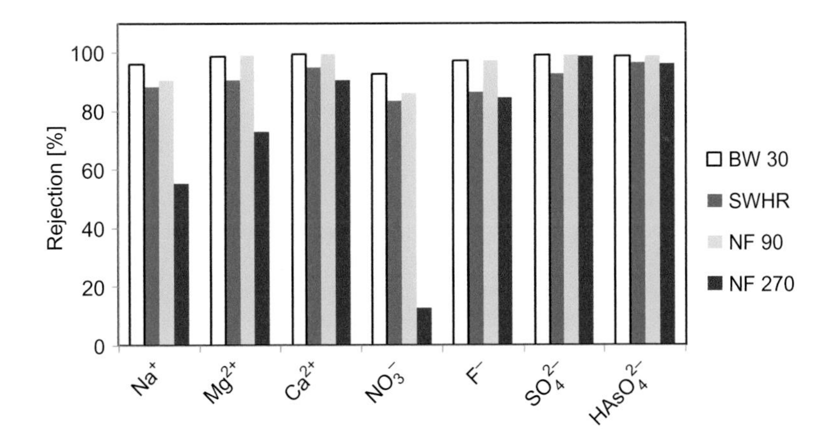

Figure 4.11. Removal of individual ions by BW30, SWHR, NF90, and NF270 membranes (Ca^{2+}: 180 mg L^{-1}, Mg^{2+}: 62 mg L^{-1}, Na^+: 140 mg L^{-1}, SO_4^{2-}: 385 mg L^{-1}, $HAsO_4^{2-}$: 0.2 mg L^{-1}, F^-: 10 mg L^{-1} and NO_3^-: 220 mg L^{-1}; Pressure, NF = 0.8 MPa, RO = 1.8 MPa; pH 7; Temperature 20°C).

cations (Na^+) was observed due to their larger hydrated radii (steric effect) (Table 4.4). Similarly, a higher rejection of divalent anions ($HAsO_4^{2-}$ and SO_4^{2-}) was obtained due to combined steric effects and electrostatical effects (Table 4.4). SO_4^{2-} ions have the largest hydrated radii, thus, the rejection was highest among all anions. Althoughthe hydrated radius of $HAsO_4^{2-}$ is smaller than that of F^- and NO_3^-, the rejection was high due to the low diffusion coefficient and high charge density (Akin *et al.*, 2011; Hoinkis *et al.*, 2011).

For nanofiltration, the rejection of ions with NF 270 was the lowest (55.4% Na^+, 72.9% Mg^{2+}, 90.4% Ca^{2+}, 13.0% NO_3^-, 84.5% F^-, 98.8% SO_4^{2-} and 96.1% $HAsO_4^{2-}$). This can be explained by the large pore size of NF 270 (~0.84 nm) (Nghiem and Hawkes, 2001) compared to NF90 (0.68 nm) (Nghiem and Hawkes, 2007). The metal ions shielded the negative charge of the membrane, which simultaneously decreased the charge of the membrane, especially when high concentrations of divalent cations were present in the solution (Choi *et al.*, 2001) and thus, the anions easily passed through the larger pore size membrane and the rejection decreased. Moreover, it was observed that the rejection of monovalent anions (NO_3^-, F^-) was lower than for divalent anions (SO_4^{2-}, $HAsO_4^{2-}$). In addition, NO_3^- ions have a lower hydrated radius and can pass through the membrane more easily than F^- (Fig. 4.11). Although the divalent cations Ca^{2+} and Mg^{2+} were highly rejected by negatively charged membranes, a small amount of divalent cations could still pass to the permeate stream, which made the monovalent anions (NO_3^-, F^-) and a small amount of $HAsO_4^{2-}$ forced to pass through the membrane for balancing the electroneutrality. Thus, a lower rejection of monovalent anions was obtained (Choi *et al.*, 2001).

Previous studies (Akin *et al.*, 2011; Teychene *et al.*, 2013) have shown a higher rejection of F^- by SWHR than by BW 30. However, in our study we found that the removal of all the ions was higher in case of BW 30 rather than SWHR. This can be explained due to the higher negative zeta potential (ZP) of the BW 30 membrane (−13 mV, pH 7) (Ishida *et al.*, 2005) than that of SWHR (−5 mV, pH 7) (Lee *et al.*, 2010) and as a result, the separation increases due to greater electrostatic repulsion between the anions and the membrane.

In summary, from the above experiments, it can be concluded that the BW30 membrane has a higher rejection of both As and F from synthetic groundwater solution (>90%) than all other membranes tested. NF90, NF270 and SWHR membranes can also be used potentially for removal of both As and F^- and other ions (except nitrate for NF270).

4.6 CONCLUSION

Several remediation technologies have been considered for removal of As, F and U from water. RO appears to be the most effective processes to remove these contaminants from aqueous solution. Thus, an application of this technique is reviewed in this chapter along with experimental observations to support the reported performances.

It is clear from the literature review that the rejection efficiency for As(V) is remarkably higher than As(III) by both high pressure and low pressure RO membranes due to electrostatic repulsion and size exclusion. Therefore, oxidation from As(III) to As(V) becomes a very important pretreatment step for removing total As below the MCL. Removal of fluoride by both high pressure and low pressure RO technologies is also very effective and are sufficient to reduce the permeate concentration below the MCL. Although there are very few studies that have evaluated removal of U from drinking water by RO membranes the removal efficiency of U have showed successful application of the technique in the drinking water industry.

Many researchers have used RO for desalination and production of potable water but there are some drawbacks, which need to be considered. Fouling on the semi-permeable RO membrane is one of the important drawbacks. In this case cleaning or replacement of the membrane or an increase in operating pressure is necessary to avoid the problem, which simultaneously increases the cost of operation. Moreover, this technology is not sufficient to lower the contaminant concentration below the MCL especially when the contaminated water has a high concentration of the inorganic pollutant and application of integrated/hybrid technology could be a better option. Additionally the removal efficiency of these toxic inorganic ions by RO depends on some important parameters, such as the source of water, pH of the solution, membrane materials and the conditions applied during the removal process. Therefore, all of these facts need to be considered and the development of better quality membranes (with higher-fouling resistance, rejection efficiency and water flux) is necessary before the application of RO in both developed as well as newly industrialized developing countries for potable water production.

ACKNOWLEDGEMENT

We acknowledge the help of European Union (Erasmus Mundus External Cooperation Window 13) for providing a doctoral fellowship to PM.

REFERENCES

Akbari, H.R., Rashidi Mehrabadi, A. & Torabian, A. (2010) Determination of nanofiltration efficiency in arsenic removal from drinking water. *Iranian Journal of Environmental Health Science & Engineering*, 7 (3), 273–278.

Akin, I., Arslan, G., Tor, A., Cengeloglu, Y. & Ersoz, M. (2011) Removal of arsenate [As(V)] and arsenite [As(III)] from water by SWHR and BW-30 reverse osmosis. *Desalination*, 281 (17), 88–92.

Atkins, P.W. (1998) *Química-física*. Ediciones Omega, Barcelona, Spain.

Baker, R.W. (2004) *Membrane technology and applications*. John Wiley & Sons, Menlo Park, CA.

Barton, C.S., Stewart, D.I., Morris, K. & Bryant, D.E. (2004) Performance of three resin-based materials for treating uranium contaminated groundwater within a PRB. *Journal of Hazardous Materials*, 116 (3), 191–204.

Bhattacharya, P., Hoinkis, J. & Figoli, A. (2009) Drinking water – sources, sanitation and safeguarding. The Swedish Research Council Formas, Stockholm, Sweden. pp. 69–91.

Cadotte, J.E. (1977) Reverse osmosis membrane. US Patent 4,039,440 (August, 1977).

Cadotte, J.E. (1981) Interfacially synthesized reverse osmosis membrane. US Patent 4,277,344 (July, 1981).

Choi, S. Yun, Z., Hong, S. & Ahn, K. (2001) The effect of co-existing ions and surface characteristics of nanomembranes on the removal of nitrate and fluoride. *Desalination*, 133 (1), 53–64.

Coleman, S.J., Coronado, P.R., Maxwell, R.S. & Reynolds, J.G. (2003) Granulated activated carbon modified with hydrophobic silica aerogel: potential composite materials for the removal of uranium from aqueous solutions. *Environmental Science & Technology*, 37 (10), 2286–2290.

Deowan, A.S., Hoinkis, J. & Pätzold, C. (2008) Low-energy reverse osmosis membranes for arsenic removal from groundwater. In: Battacharya, P., Ramanathan, A.L., Bundschuh, J., Keshari, A.K. & Chandrasekharam, D. (eds.) *Groundwater for sustainable development problems, perspectives and challenges*. CRC Press, Boca Raton, FL. pp. 275–386.

Diawara, C.K., Diop, S.N., Diallo, M.A., Farcy, M. & Deratani, A. (2011) Performance of nanofiltration (NF) and low pressure reverse osmosis (LPRO) membranes in the removal of fluorine and salinity from brackish drinking water. *Journal of Water Resource and Protection*, 3 (12), 912–917.

Dolar, D., Košutić, K. & Vučić, B. (2011) RO/NF treatment of wastewater from fertilizer factory – removal of fluoride and phosphate. *Desalination*, 265 (1–3), 237–241.

Endoh, R.T., Kurihara, M. & Ikeda, K. (1977) New polymeric materials for reverse osmosis membranes. *Desalination*, 21 (1), 35–44.

Figoli, A., Cassano, A., Criscuoli, A., Salatul Islam Mozumder, M., Tamez Uddin, A., Akhtarul Islam, M. & Drioli, E. (2010) Influence of operating parameters on the arsenic removal by nanofiltration. *Water Research*, 44 (1), 97–104.

Floch, J. & Hideg, M. (2004) Application of ZW-1000 membranes for arsenic removal from water sources. *Desalination*, 162 (10), 75–83.

Gafvert, T., Ellmark, C. & Holm, E. (2002) Removal of radionuclides at a waterworks. *Journal of Environmental Radioactivity*, 63 (2), 105–115.

Gamal Khedr, M. (2013) Radioactive contamination of groundwater, special aspects and advantages of removal by reverse osmosis and nanofiltration. *Desalination*, 321 (1), 47–54.

Geucke, T., Deowan, S., Hoinkis, J. & Pätzold, C. (2009) Performance of a small-scale RO desalinator for arsenic removal. *Desalination*, 239 (1–3), 198–206.

Gholami, M.M., Mokhtari, M.A., Aameri, A. & Alizadeh Fard, M.R. (2006) Application of reverse osmosis technology for arsenic removal from drinking water. *Desalination* 200, 725–727.

Hoinkis, J., Valero-Freitag, S., Caporgno, M.P. & Pätzold, C. (2011) Removal of nitrate and fluoride by nanofiltration – a comparative study. *Desalination and Water Treatment*, 30 (1–3), 278–288.

Horvath, A.G. (1931) Water softening. US Patent 1,825,631 (September, 1931).

Ishida, K.P., Bold, R.M. & Phipps, D.W., Jr. (2005) Identification and evaluation of unique chemicals for optimum membrane compatibility and improved cleaning efficiency. Orange County Water District (ed.), Fountain Valley, CA.

Katadyn Power Survivor 160 E. Owner's manual. Available from: katadynch.vs31.snowflakehosting.ch/fileadmin/user_upload/katadyn_products/Downloas/Manual_Katadyn_PS-160E_EN.pdf [accessed June 2009].

Kryvoruchko, A.P., Yurlova, L.Y., Atamanenko, I.D. & Kornilovich, B.Y. (2004) Ultrafiltration removal of U(VI) from contaminated water. *Desalination*, 162 (10), 229–236.

Lee, W., Ahn, C.H., Hong, S., Kim, S., Lee, S., Baeka, Y. & Yoon, J. (2010) Evaluation of surface properties of reverse osmosis membranes on the initial biofouling stages under no filtration condition. *Journal of Membrane Science*, 351 (1–2), 112–122.

Linde, K. & Jönson, A.S. (1995) Nanofiltration of salt solutions and landfill leachate. *Desalination*, 103 (3), 223–232.

Loeb, S. & Sourirajan, S. (1963) Sea water demineralization by means of an osmotic membrane, in saline water conversion II. In: Gould, R.F. (ed.) *Advances in Chemistry Series*, Number 38. American Chemical Society, Washington, DC. pp. 117–132.

McKinney, R. & Rhodes, J.H. (1971) Aromatic polyamide membranes for reverse osmosis separations. *Macromolecules*, 4 (5), 633–637.

Mondal, P., Bhowmick, S., Chatterjee, D., Figoli, A. & Van der Bruggen, B. (2013) Remediation of inorganic arsenic in groundwater for safe water supply: a critical assessment of technological solutions. *Chemosphere*, 92 (2), 157–170.

Montaña, M., Camacho, A., Serrano, I., Devesa, R., Matia, L. & Vallés, I. (2013) Removal of radionuclides in drinking water by membrane treatment using ultrafiltration, reverse osmosis and electrodialysis reversal. *Journal of Environmental Radioactivity*, 125 (1), 86–92.

Nghiem, L.D. & Hawkes, S. (2007) Effects of membrane fouling on the nanofiltration of pharmaceutically active compounds (PhACs): mechanisms and role of membrane pore size. *Separation and Purification Technology*, 57 (1), 182–190.

Nightingale Jr., E.R. (1995) Phenomenological theory of ion solvation. Effective radii of hydrated ions. *The Journal of Physical Chemistry*, 63 (9), 1381–1387.

Perera, D.H.N., Nataraj, S.K., Thomson, N.M., Sepe, A., Hüttner, S., Steiner, U., Qiblawey, H. & Sivaniah, E. (2014) Room-temperature development of thin film composite reverse osmosis membranes from cellulose acetate with antibacterial properties. *Journal of Membrane Science*, 453 (1), 212–220.

Reid, C.E. & Breton, E.J. (1959) Water and ion flow across cellulosic membranes. *Journal of Applied Polymer Science*, 1 (2), 133–143.

Richards, L.A., Richards, B.S., Rossiter, H.M.A. & Schäfer, A.I. (2009) Impact of speciation on fluoride, arsenic and magnesium retention by nanofiltration/reverse osmosis in remote Australian communities. *Desalination*, 248 (1–3), 177–183.

Richards, L.A., Vuachère, M. & Schäfer, A.I. (2010) Impact of pH on the removal of fluoride, nitrate and boron by nanofiltration/reverse osmosis. *Desalination*, 261 (3), 331–337.

Robinson, R.A. & Stokes, R.H. (eds.) (1965) *Electrolyte solutions*. Butterworths, London, UK.

Rossiter, H.M.A., Graham, M.C. & Schäfer, A.I. (2010) Impact of speciation on behaviour of uranium in a solar powered membrane system for treatment of brackish groundwater. *Separation and Purification Technology*, 71 (1), 89–96.

Saitua, H., Gil, R. & Perez Padilla, A. (2011) Experimental investigation on arsenic removal with a nanofiltration pilot plant from naturally contaminated groundwater. *Desalination*, 274 (1–3), 1–6.

Sehn, P. (2008) Fluoride removal with extra low energy reverse osmosis membranes: three years of large scale field experience in Finland. *Desalination*, 223 (1–3), 73–84.

Shuibo, X., Chun, Z., Xinghuo, Z., Jing, Y., Xiaojian, Z. & Jingsong, W. (2009) Removal of uranium (VI) from aqueous solution by adsorption of hematite. *Journal of Environmental Radioactivity*, 100, 162–166.

Singh, A.K., Bhagowati, S., Das, T.K., Yubbe, D., Rahman, B., Nath, M., Obing, P., Singh, W.S.K., Renthlei, C.Z., Pachuau, L. & Thakur, R. (2008) Assessment of arsenic, fluoride, iron, nitrate and heavy metals in drinking water of northeastern India. *ENVIS Bulletin: Himalayan Ecology*, 16 (1), 1–7.

Strathmann, H. (1990) Synthetic membranes and their preparation. In: Porter, M. (ed.) *Handbook of industrial membrane technology*. Noyes, Park Ridge, NJ. pp. 1–60.

Teychene, B., Collet, G., Gallard, H. & Croue, J.P. (2013) A comparative study of boron and arsenic (III) rejection from brackish water by reverse osmosis membranes. *Desalination*, 310, 109–114.

Velizarov, S., Crespo, J. & Reis, M. (2004) Removal of inorganic anions from drinking water supplies by membrane bio/processes. *Reviews in Environmental Science and Biotechnology*, 3 (4), 361–380.

Vos Jr., K.D., Burris, F.O. & Riley, R.L. (1966) Kinetic study of the hydrolysis of cellulose acetate in the pH range of 2–10. *Journal of Applied Polymer Science*, 10 (5), 825–832.

Vrijenhoek, E.M. & Waypa, J.J. (2000) Arsenic removal from drinking water by a "loose" nanofiltration membrane. *Desalination*, 130 (3), 265–277.
Wang, D., Su, M., Yu, Z., Wang, X., Ando, M. & Shintani, T. (2005) Separation performance of a nanofiltration membrane influenced by species and concentration of ions. *Desalination*, 175 (2), 219–225.
Wang, L.K., Chen, J.P., Hung, Y.T. & Shammas, N.K. (2011) Membrane and desalination technology. Volume 13 of the *Handbook of environmental engineering*. Springer, New York, Dordrecht, Heidelberg, London.
Yoon, J., Amy, G., Chung, J., Sohn, J. & Yoon, Y. (2009) Removal of toxic ions (chromate, arsenate, and perchlorate) using reverse osmosis, nanofiltration, and ultrafiltration membranes. *Chemosphere*, 77 (2), 228–235.

CHAPTER 5

Electro-membrane processes for the removal of trace toxic metal ions from water

Svetlozar Velizarov, Adrian Oehmen, Maria Reis & João Crespo

5.1 INTRODUCTION

The contamination of water sources with toxic metals and semi-metallic elements, including arsenic (As) and selenium (Se), is a matter of great concern worldwide, because of their potential negative impact on the ecosystems. While in very small amounts, many of these metals are necessary to support life, in larger amounts, they can become extremely toxic. Their bioaccumulation in animals and human bodies may lead to long-term negative health effects and chronic diseases (Richardson, 2003; Smith *et al.*, 2002). Aqueous streams, containing toxic metals (Cd, Ni, Hg, Pb, Co, Cu, Zn, Al, Sb, Mo, Sn, V, U, etc.) are produced in many industrial processes, such as metal-finishing applications (e.g., electroplating), production of accumulators and batteries, fuel, paints, pesticides and cellulose acetate manufacturing, etc. Due to intensive mining activities, the water in the vicinity of such places can become severely polluted. Therefore, the maximum allowed concentrations of such compounds are generally set by the drinking water quality regulatory standards in the relatively low $\mu g\,L^{-1}$ to $mg\,L^{-1}$ range, therefore, the majority of them can be referred to as trace metal pollutants or micropollutants. Since, usually there are no detectable organoleptic changes in drinking water in the presence of toxic metal ions in trace levels, it is rather possible that some of them may easily remain undetected, thus additionally increasing the possible health risks. Therefore, environmental sustainability requires a complete removal of these contaminants from the water cycle.

Membrane separation processes, if properly selected, offer the advantage of producing high quality drinking water. In many cases, one membrane process can be followed by another or applied in combination of physical, chemical and/or biological processing, to produce water of even higher quality. In these processes, the membrane can be viewed as a barrier between contaminated and purified water streams. This physical phase separation of the two streams often allows for operation with no or minimal chemical water pre-treatment, which otherwise forms deleterious by-products (Bergman *et al.*, 1995; Jacangelo *et al.*, 1997). Pressure-driven membrane processes such as reverse osmosis (RO) and nanofiltration (NF) are well-developed and widely used technologies for water treatment (Shih, 2005; Van der Bruggen *et al.*, 2008). However, they are wasteful because they require most of the water to be permeated through the membrane and all or most solutes are retained instead of only the target trace contaminant(s).

Since most metal-containing species in water are either positively (cations) or negatively (anions) charged, the use of electro-membrane processing for their removal appears as a natural choice. Therefore it is not surprising that the scientific and patent literature devoted to possible applications of electro-membrane processing for removal of trace toxic metal ions from water is abundant. Therefore this chapter does not pretend to be exhaustive and covers mainly some recently published studies, focusing on hybrid processes, which are gaining increasing attention. Older studies that have explored possible applications of electromembrane processes in water treatment have been discussed elsewhere (e.g., Banasiak, 2009; Davis *et al.*, 2000; Grimm *et al.*, 1998; Strathmann, 2004; Koter and Warszawski, 2000; Strathmann, 2010; Velizarov *et al.*, 2004). After briefly presenting the most relevant electromembrane processes for treating waters, containing traces of toxic metal ions, and addressing the removal of arsenic (As), fluoride (F^-) and

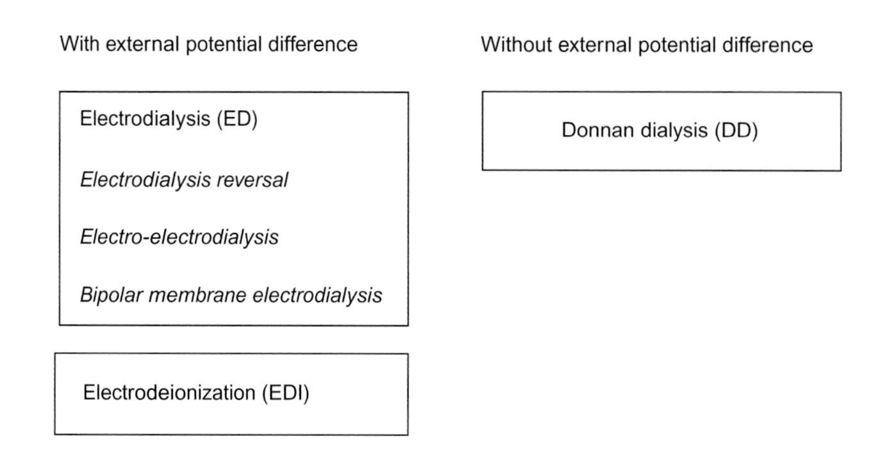

Scheme 5.1. Electro-membrane processes for separation of metal ions from water.

uranium (U), we have selected and discussed two case-studies developed in our recent research, in order to illustrate possible successful application of Donnan dialysis, applied as a single or part of an integrated treatment for the removal of arsenate and ionic mercury from contaminated water sources. They are based on a novel water treatment concept, combining a continuous membrane transport of an ionic pollutant with its simultaneous chemical or biological treatment (referred to as ion-exchange membrane reactor (IEMR) or ion-exchange membrane bioreactor (IEMB), respectively) (Velizarov *et al.*, 2011).

Membrane processes that use ion-exchange membranes and electric potential difference as the driving force for ionic species transport are referred to as electromembrane processes (Strathmann, 2004). The following electro-membrane separation processes (Scheme 5.1) can be distinguished: electrodialysis (ED), including variations such as electrodialysis reversal, electro-electrodialysis and bipolar membrane electrodialysis, electrodeionization (EDI), and Donnan dialysis (DD).

Although the driving force in Donnan dialysis is not an external electric potential difference but a concentration difference, the latter leads to the establishment of an internal electric (Donnan) potential difference, which can be utilized for transport and separation of target ionic species. Moreover, as will be discussed in more detail later, this process is especially appropriate for removing trace target ions from low salinity waters. The Donnan dialysis type of operation requires the presence in the stripping solution of a so-called "driving" counter-ion, which is transported across the membrane in a direction opposite to that of the target counter-ion(s) in order to maintain overall electroneutrality in the system. From the operational, economic and environmental points of view, chloride has been generally considered as a suitable driving counter-ion when anionic pollutants have to be removed from contaminated drinking water supplies (Velizarov *et al.*, 2004). In the case of cationic pollutants, sodium, potassium or hydrogen ions appear to be the best possible choices.

In ED, the transport of ions present in contaminated water is accelerated due to an electric potential difference applied externally by means of electrodes (anode and cathode). In this process, anion-exchange membranes and cation-exchange membranes are applied in order to transport anions and cations to the anode or cathode, respectively. Process deterioration due to membrane scaling is a frequently observed problem; therefore, the ED systems are usually operated in the so-called electrodialysis reversal mode, in which the polarity of the electrodes is reversed several times per hour to change the direction of ion movement (Strathmann, 2004). The external electric potential driving force allows higher ionic fluxes to be obtained than those achievable in DD, but a different degree of demineralization (desired anions and/or cations are also removed from the water) depending on the voltage and type of the membranes used is obtained.

Electrodeionization (EDI) is a hybrid process combining ion-exchange with electrodialysis by introducing ion-exchange resins into the electrodialysis chambers. The combination allows for treating very dilute electrolyte solutions, while the ion-exchange resin beads inside the chambers are continuously regenerated *in-situ* by hydrogen and hydroxide ions produced by water electrolysis occurring in the two external electrode compartments (Monzie *et al.*, 2005). Therefore, the EDI process has received increasing attention in the purification of solutions containing toxic metal ions (Dzyako and Belyakov, 2004; Grebenyuk *et al.*, 1998; Mahmoud and Hoadley, 2012; Spoor *et al.*, 2002a; 2002b).

One of the major possible drawbacks is that EDI is susceptible to precipitation of bivalent metal hydroxides as a result of metal ions reacting with hydroxide ions generated within the EDI apparatus. The implementation of EDI in water treatment has been greatly limited by this drawback (Feng *et al.*, 2007). Therefore, possible integrations of ED with other processes (other than EDI) are emerging (Abou-Shady *et al.*, 2012; Nataraj *et al.*, 2007; Tran *et al.*, 2012).

When the purpose is toxic metal(s) removal, reduction in water hardness could be a desired side effect in some cases but in others may cause too "deep" softening (as in RO treatment), therefore the applicability of ED and EDI depends strongly on the polluted water ionic composition (Velizarov *et al.*, 2004).

In summary, electromembrane processes, especially when used as part of hybrid treatment schemes, can provide an efficient removal of toxic metal ions from water. When a target metal exists in water as a mono-valent species, the use of mono-valent (cation- and/or anion-permselective membranes) is especially attractive. Situations, in which ED appears to be less applicable are for waters of very low salinity (conductivity of less than 0.5 mS cm^{-1}), for which EDI or DD can be better choices, and, in cases when besides ions, removal of low molecular mass non-charged compounds from the water is desired. In the latter case, pressure-driven membrane processes such as RO or NF may be preferable.

5.1.1 *Removal of uranium fluoride and arsenic by electro-membrane processes*

Uranium is a very toxic and radioactive heavy metal found in nuclear effluents also naturally and in uranium, coal, hydrocarbon exploitation and associated activities. It is therefore not surprising that it was the first to be historically considered for electro-membrane treatment (Davis *et al.*, 1971; Wallace, 1967).

The feasibility of applying electro-membrane processes as DD, ED and EDI has been evaluated. Since concentration ratios determine the Donnan equilibrium, not concentration differences, DD allows for transport of a target counter-ion against its own concentration gradient, it is a convenient method for treating water containing only trace levels of toxic ions. Furthermore, the hydraulic residence time can be independently adjusted in the two compartments (feed and stripping), thus allowing the degree of extraction of the target toxic pollutant to be optimized. Due to these characteristics, the removal of U (in the form of uranuyl ions $(UO_2)^{2+}$) from water streams by Donnan dialysis has received a lot of attention.

Cation-exchange membranes were assembled in a plate-and-frame Donnan dialyzer with solution compartments arranged so that the feed and stripping streams flow in a countercurrent mode of operation. This apparatus was used to recover $UO_2(NO_3)_2$ from a 0.01 M feed to a final content of 0.28 M with 2 M HNO_3 as the stripping electrolyte and to 0.46 M with 2 M H_2SO_4 as the stripping electrolyte (Wallace, 1967).

Although DD allows for achieving very high degrees of separation and concentration of target counter-ions, its possible limitation is the relatively slow transport kinetics because of the absence of externally applied electric potential difference. Therefore ED has been also studied for $(UO_2)^{2+}$ removal (Zaki, 2002). Indeed, applying a potential difference of 30 V across the used Nafion membranes, and by using Na_2CO_3 as the stripping electrolyte solution, the membrane flux of $(UO_2)^{2+}$ was increased by two orders of magnitude. However, this was at the price of a very significant electroosmotic water transport, which can be considered as an undesirable side effect.

More recently, ED and EDI have been applied for treatment of dilute U-containing synthetic aqueous streams. The use of ED was found to be effective, but the presence of magnesium ions in the feed solution caused a decrease in U removal by the subsequent "polishing" EDI process (Zaheri *et al.*, 2010).

Besides treatment of nuclear plant effluents, a more recent trend in the field of membrane-assisted U removal is the treatment of water streams possessing natural radioactivity. A pilot ED plant has been built to test the possibility to improve the quality of the water supplied to the Barcelona metropolitan area from the Llobregat River through reducing its natural radioactivity. The results obtained revealed a significant improvement in the radiological water quality provided by ED with removal rates higher than 60% for gross alpha, gross beta or U activities (Montana *et al.*, 2013).

In the 1990s, the F^- drinking water sources contamination problem, which is especially important in some African countries, stimulated research on possible ways of its removal to the desired low concentration. It is important to note that, contrary to the cases of U and As, low levels ($\sim$1 mg L^{-1}) of F^- in the drinking water are health beneficial.

Fluoride removal by Donnan dialysis was investigated and mathematically modeled for a bi-ionic system (NaF as the feed and NaCl as the strip) by Dieye *et al.* (1998) and later on tested in synthetic drinking waters with compositions close to those found in some natural waters in Africa (Hichour *et al.*, 2000). An F^- concentration in agreement with the norm (<1.5 mg L^{-1}) was reached in the later study and the addition of a complex-forming ion such as Al^{3+} (to obtain F^- complexes, which are not able to cross the membrane) to the strip solution allowed a low free F^- concentration in the strip and a reasonably high process driving force to be maintained. A pilot-scale plate-and-frame DD module, consisting of eleven cells (five feed and six strip cells) separated by DSV anion-exchange membranes (Asahi Glass was tested and the maximum treated water production rate reached 2.5 L m^{-2} h^{-1}); however, the treated water salinity was increased by about 25% due to electrolyte leakage from the strip to the feed solution. It may be concluded that the Donnan co-ion exclusion provided by this type of membrane was not sufficiently high.

The use of a Neosepta ACS membrane (Tokuyama) was able to solve this problem in a subsequent study by Garmes *et al.* (2002), who also investigated a combination of a Donnan dialytic transport of F^- with its adsorption on Al_2O_3 or ZrO_2, added to the stripping solution.

ED has been also extensively studied as a possible way for removing F^- from water. Using model NaF salt solutions, it has been reported that ED is most effective only if F^- is either the single anionic specie in the water or is present in great excess (Kabay *et al.*, 2008). This situation is, however, rather unlikely to occur in natural waters, which as a rule contain various ionic species at higher levels than the F^- level. Especially problematic is chloride, because due to the anion-exchange membrane resin preference for Cl^- over F^-, any presence of chloride would reduce the ED process efficiency for F^- removal from water streams.

Brackish water, containing 3000 mg L^{-1} of TDS and 3 mg L^{-1} of F^-, was tested by Amor *et al.* (2001). The use of a mono-anion permselective membrane (Neosepta ACS) allowed for maintaining the water sulfate concentration close to its original value (only 5% was removed). The membrane transported the anions in the following order: $Cl^- > F^- > HCO_3^- > SO_4^{2-}$.

Using the same Neosepta ACS and a Neosepta CMX as the cation-exchange membrane, an ED process for testing the feasibility for removing F^- from natural groundwater in Morocco has been also tested. It was found that the required F^- drinking water content (<1.5 mg L^{-1}) can be obtained, but at a very high overall water demineralization rate of 80% (Sahli *et al.*, 2007).

Therefore it can be concluded that ED treatment of multi-ionic water with a relatively high F^- concentration for defluoridation appears to be not feasible, because it would require a very high amount of energy. Moreover, concentration polarization problems at the membrane surfaces can occur in both dilute and concentrate compartments.

Arsenic (As) removal from drinking water supplies by electro-membrane processes until recently has been rarely reported. The latter may stem from the fact that the required target As value in the treated water must be extremely low (10 μg L^{-1} as As). Therefore it could be anticipated that the accompanying anions competition and water demineralization problems, especially in the case of ED, would be much more severe compared to the case of F^- removal (required to the very low F^- mg L^{-1} instead of μg L^{-1} range). In any case, during the last few years, there is an obvious increasing interest in this topic, which has led to gaining a much better insight into the mechanisms of arsenate/arsenite transport across anion-exchange membranes (Guell *et al.*,

2011; Velizarov *et al.*, 2005; 2013; Zhao *et al.*, 2010) as well as to testing a DD-based pilot plant (Zhao *et al.*, 2012).

Since it is widely distributed throughout the earth's crust, As represents one of the most serious environmental concerns worldwide. Although organic forms of As are possible in water, only the inorganic arsenite (As(III)) and arsenate (As(V)) forms have been found to be significant in groundwater (Henke *et al.*, 2009). The maximum allowed concentration of this element in drinking water is set to $10\,\mu g\,L^{-1}$ (USEPA, 2001; WHO, 2008). In Portugal, recent findings have shown high levels in numerous locations (located far from centralized drinking water facilities and serving populations in remote rural locations), in some cases exceeding the recommended limit by more than fifty times (Garcia, 2006).

While arsenate removal by conventional anion-exchange has been extensively explored, only a few studies have been reported on the separation of As-containing ions by Donnan dialysis (Guell *et al.*, 2011; Velizarov *et al.*, 2005; 2013; Zhao *et al.*, 2010; 2012). Unlike conventional ion-exchange, involving resin loading and regeneration steps, Donnan dialysis is an ion-exchange membrane based separation process that can be performed under either batch or continuous operation conditions (Davis, 2000). Another potential advantage of applying Donnan dialysis instead of conventional ion-exchange, is that the degree of removal of a target charged pollutant can be optimized through independently adjusting the hydraulic residence times in the feed (contaminated water) and stripping compartments that are separated by the ion-exchange membrane.

Zhao *et al.* (2010) performed batch Donnan dialysis studies at different pH values using two types of anion-exchange membranes, one homogeneous and another heterogeneous. They reported that the arsenate removal efficiency was higher for the case of the homogenous membrane. Furthermore, it has been recently revealed that the use of membranes with mono-anion-permselective properties such as Neosepta ACS is not recommended because of slow transport of arsenate and its significant retention in the membrane phase (Velizarov, 2013).

5.1.2 *The ion-exchange membrane bioreactor concept*

A novel hybrid process concept, referred to as ion-exchange membrane bioreactor (IEMB) for the transport of ionic pollutants present in water through an ion-exchange membrane and their simultaneous biotransformation to harmless products, has been proposed by Crespo and Reis (2001). More recently, this concept was tested for the simultaneous transport and chemical precipitation of arsenate and the designation of the process was modified to ion-exchange membrane reactor (IEMR) in order to highlight the absence of biological treatment in the latter case (Oehmen *et al.*, 2011).

The IEMB concept integrates the transport of target ionic pollutants from an aqueous stream, through a dense ion-exchange membrane, to a stripping compartment where these ionic compounds are biologically reduced to harmless products (see Fig. 5.1). In order to enhance the transport of the target ionic pollutants, an appropriate counter ion is added to the stripping compartment at sufficiently high concentration. This procedure makes it possible to transport the target pollutant, even against its own concentration gradient. Once the ionic pollutant reaches the stripping compartment, it is converted by an appropriate microbial culture, which is able to reduce it under anoxic conditions. A non-charged and non-fermenting carbon source (used as electron donor, e.g., ethanol) is fed to the stripping biological compartment in enough amounts, in order to assure the complete biological reduction of the transported ionic solutes.

This concept is rather simple but it presents a series of relevant features, which answer to problems previously encountered in physical/chemical and biological treatment of charged pollutants:

- The microbial culture able to reduce the target pollutant(s) is physically separated from the water stream by a dense, non-porous membrane, thus assuring that the treated water is never in contact with the microorganisms responsible for the bioconversion.

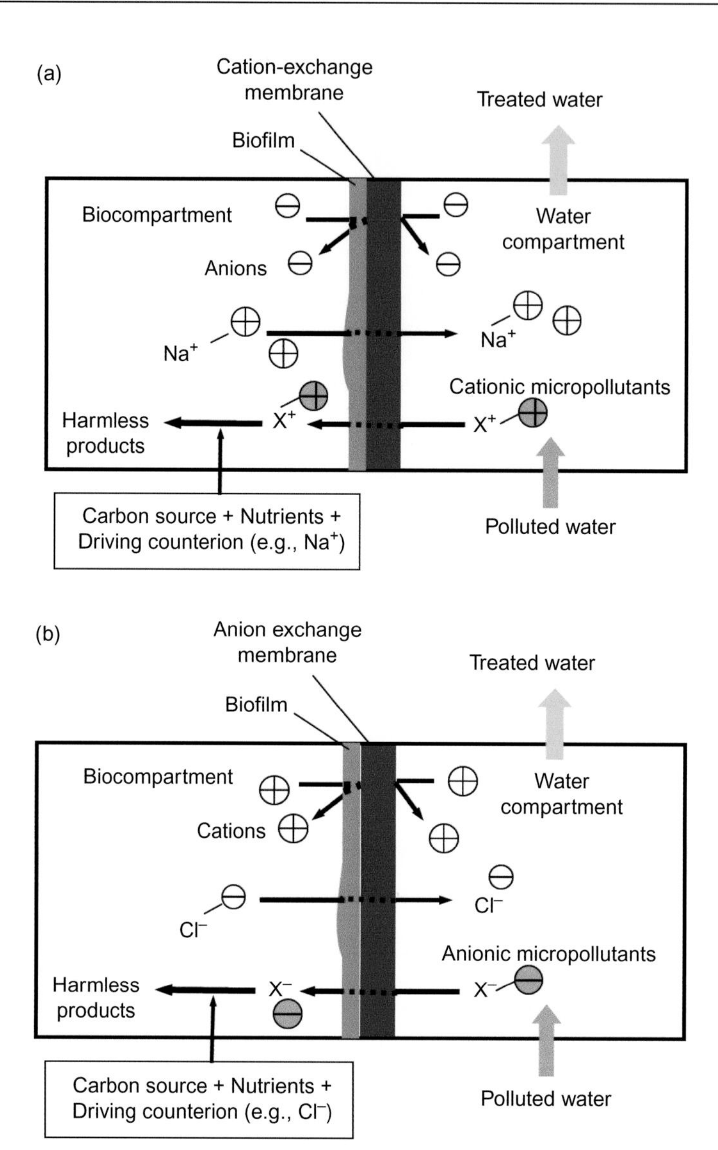

Figure 5.1. Schematic diagram of the ion transport mechanism in the ion-exchange membrane bioreactor (IEMB). (a): for removal of cationic and (b): anionic micropolutants.

- As the membrane selected is charged (fixed charge opposite to the charge of the solutes aimed to be transported) and non-porous, it is possible to assure that transport of non-charged metabolic by-products is extremely reduced; if they present a charge identical to that of the membrane fixed functional sites, their transport may be completely avoided.
- Naturally, a biofilm develops on the membrane surface contacting the biological compartment. This biofilm acts as an active reaction zone, where most of the aimed reaction process takes place, also providing an additional barrier to the transport of excess carbon source, used as electron donor, from the biomedium to the treated water. These three features assure that secondary contamination of treated water by microorganisms, metabolic by-products and excess carbon source can be avoided if an appropriate membrane and operating conditions are selected. Also, strict control of the rate of addition of carbon source, aiming to avoid a situation of excess or deprivation, is not required.

- As the microbial culture can be selected by using appropriate selective pressure conditions, only target ionic pollutants are converted. As a consequence, the driving force for transport of ionic compounds from the water stream is only kept high for compounds that are biologically converted. This feature assures that the water stream is not unspecifically depleted of ions which may be important to maintain an adequate water composition balance (in opposition to some physical water treatment processes, such as reverse osmosis).
- As anoxic conditions are used in the biological compartment, the yield (and rate) of cell mass production is much lower than the one typical for aerobic membrane bioreactors, where biofilm development represents a significant problem in terms of resistance to mass transfer and membrane clogging and fouling. In the IEMB system, the biofilms that develop naturally at the membrane surface are rather thin, having a positive impact as an active reaction zone and barrier to carbon source loss through the membrane. Also, as the membrane is non-porous, clogging problems are not an issue. The IEMB system proved that it can be operated during long periods of time without flux decline or need for membrane cleaning (tests with an extension of 4 months have been accomplished).
- The concentration of driving counter-ion (e.g., Cl^-) may be adjusted in order to assure a high transport rate of the target polluting ion, even against its own concentration gradient. This feature is also particularly interesting if (e.g., due to some operating problem) the target ion accumulates in the biocompartment.
- As the target ionic pollutants are converted to harmless products (e.g., N_2, Cl^-) and not just transferred and concentrated in a stripping compartment/stream, as happens in physical processes for water treatment, brine solutions are not produced, thus avoiding the need of their treatment and disposal.
- The hydraulic residence time (HRT) can be adjusted independently in both water and biological compartments. Typically, very large residence times may be used in the biological compartment (the biological reaction takes mostly place within the biofilm at the surface of the membrane), originating a waste stream with an extremely reduced volume when compared with the volume of treated water.

The treated water throughput of the IEMB system depends strongly from the driving force for transport of the target polluting anion. For relatively low, but common, levels of contamination the IEMB system allows the micropollutant concentration in the treated water to be reduced to target values, at a throughput of 30–10 L/(m^2 of membrane × h). This throughput is rather competitive, even when compared with common throughputs obtained by nanofiltration and/or reverse osmosis. Additionally, as mentioned before, the IEMB presents the advantages of leading to a well-balanced treated water (in terms of its ionic composition) and not forming a brine stream that requires further disposal/treatment (which may be costly, according to local dispositions).

The IEMB process has shown excellent performance for the case of removal of monovalent anions, such as bromate (Matos *et al.*, 2008) as well as nitrate and perchlorate (Ricardo *et al.*, 2012). On the other hand, while the removal of some ions (nitrate, nitrite, perchlorate, and bromate) is relatively easier, further research is needed to extend this approach to the more challenging cases of metal-containing ions such as arsenate, chromate, ionic mercury, etc. These situations are more complex due to the strong pH dependence of their speciation in water, affecting their charge, solubility and complex-forming behavior (DeZuane, 1997).

5.2 CASE STUDIES

5.2.1 *Arsenic removal from groundwater by a hybrid DD-coagulation process*

Contrary to the cases of nitrate, perchlorate and bromate, neither the biologically catalyzed reduction of arsenate to arsenite nor the oxidation of arsenite to arsenate are beneficial since they do not result in the formation of innocuous products. Since the As-related toxicity is due to the metal itself and not due to the oxy-anionic form in which it is present, biotransformation

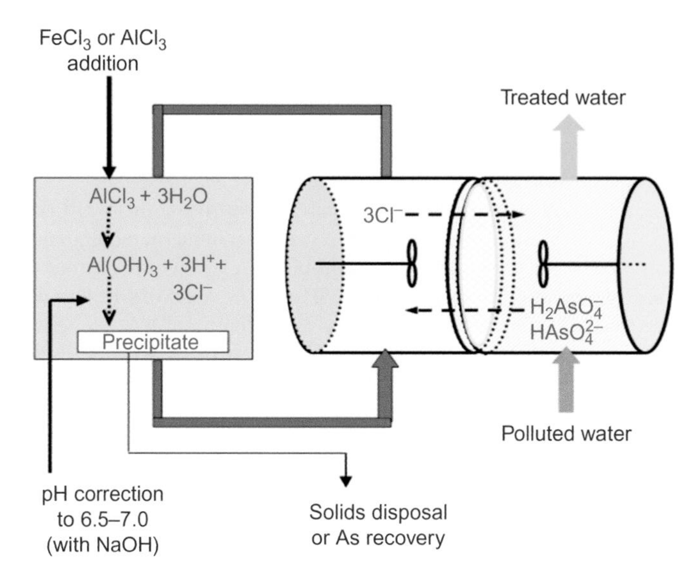

Figure 5.2. Schematic representation of the integrated concept for combined transport and treatment of arsenic species (right: polluted water entrance and outlet of treated water; left: a re-circulation vessel to which $FeCl_3$ or $AlCl_3$ is periodically added for arsenic precipitation).

is ineffective in this case. Therefore, a modified integrated process was developed (Oehmen *et al.*, 2011) through replacing the application of a mixed microbial culture in the stripping compartment by co-precipitation with iron or aluminum as a way of achieving As removal in the stripping compartment solution (Fig. 5.2). Therefore, the process was referred to as an IEMR, instead of IEMB in order to more correctly represent its features.

The approach is based on the isolation of the contaminated drinking water stream by an anion-exchange membrane barrier, through which arsenate diffuses to a stripping compartment, operated as a closed vessel with re-circulation, to which a coagulant ($FeCl_3$ or $AlCl_3$) and, if necessary, pH-controlling reagents can be periodically added to guarantee the most appropriate conditions for As precipitation. The membrane excludes the transport of cations (including Fe^{3+}, Al^{3+}) while permitting the flow of anions.

The transport of arsenate to the stripping compartment is therefore stimulated by the excess of Cl^- available for arsenate counter transport through the membrane, according to Donnan dialysis principles. The chemical precipitation additionally keeps the arsenate concentration in the stripping compartment at low levels, thus ensuring high driving force for its transport. Thus, the chemical coagulant ($FeCl_3/AlCl_3$) is completely utilized as a chemical precipitant (i.e., Fe^{3+} or Al^{3+}) and as a source of counter-ions for arsenate transport (i.e., Cl^-), simultaneously achieving both purposes with the addition of only one chemical. For very dilute electrolyte solutions, the counter-ion flux through ion-exchange membranes is diffusion boundary layer controlled, since it is directly proportional to the counter-ion concentration in water (Velizarov, 2003) An increase in the F/A ratio led to a proportional increase in the transport driving force for arsenate, and subsequently, its flux through the membrane. At high water throughputs, the arsenate residence time in the water compartment becomes very short (in the range of few minutes), which probably introduces an additional resistance to its transport due to the kinetics of the required anion-exchange reaction at the membrane – water interface, thus diminishing the arsenate flux and increasing the As concentrations in the treated water (Fig. 5.3).

The R204-UZRA membrane was selected amongst a number of anion-exchange membranes tested (Zhao *et al.*, 2011) as the most suitable membrane for the As removal process, due to the comparatively high flux of arsenate, good mechanical properties and more affordable cost. Next,

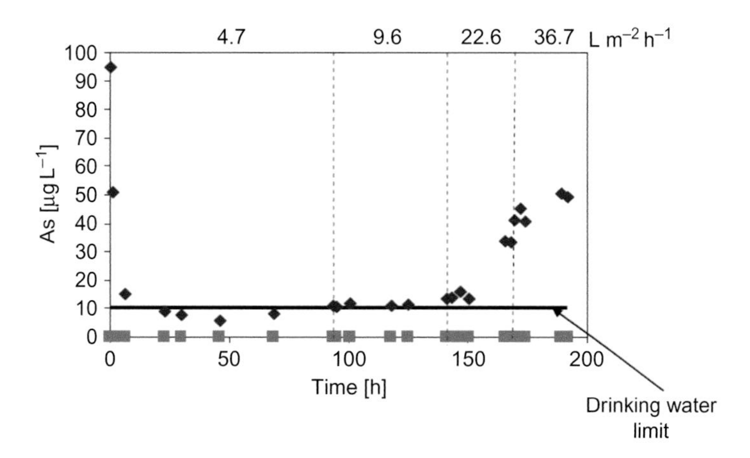

Figure 5.3. Arsenic concentration in treated water (arsenic concentration in the polluted water = 100 $\mu g\,L^{-1}$). $FeCl_3$ was used as the coagulant. The system was operated at different feed water flow rate (F) to membrane area (A) ratios (F/A = from 4.7 to 36.7 $L\,m^{-2}\,h^{-1}$) corresponding to the time intervals marked by dotted lines.

the optimal membrane to reactor volume ratios and reactor operation regimes were identified. $FeCl_3$ presented the best coagulant behavior under low flow to membrane area (F/A) ratios. When iron was added to the stripping compartment, the As concentration was maintained below the detection limit (<0.5 $\mu g\,L^{-1}$) throughout the entire experiment, The As concentration in the treated water for different polluted water flow rates was analyzed through ICP measurements and the results obtained are presented in Figure 5.3.

The results obtained showed that this integrated process can remove efficiently arsenate present in drinking water supplies, at a water treatment rate per square meter of membrane as high as 10 $L\,m^{-2}\,h^{-1}$. If higher throughputs are needed, the treated water obtained can be further blended with water from an As-free source (if available) in order to be conform with the current maximum contaminant limit of 10 $\mu g\,L^{-1}$ of As in drinking water. Such treatment rates are common for nanofiltration and demonstrate that they can be achieved also in a process that does not utilize pressure as a driving force. However, it has to be pointed out that the process was effective in avoiding secondary contamination of the treated water by the undesirable presence of coagulants, even at very high dosage levels. The latter translates into extremely high drinking water quality, which is the principal advantage of the hybrid process compared to the traditional coagulation/precipitation process.

5.2.2 *Mercury removal via the ion-exchange membrane bioreactor (IEMB)*

Mercury (Hg) is the most toxic heavy metal, and offers no beneficial biological function in any of its forms (Nies, 1999). Mercury pollution in the environment has previously triggered disasters on numerous occasions, such as the poisoning of large populations from Minnamata, Japan or in other countries such as Iraq, Brazil, Indonesia, the USA and China (Jiang, 2006). Industrial activities including mining and the chlor-alkali process have been linked with mercury contaminated water supplies, although on occasion natural sources have been responsible for high mercury levels (Lisha *et al.*, 2009). The drinking water limit for Hg recommended by the World Health Organization is 1 $\mu g\,L^{-1}$, and the by the USEPA is 2 $\mu g\,L^{-1}$. Hg levels above these limits have been previously observed in drinking water supplies (Barringer *et al.*, 2006) and reservoirs that could otherwise be used as drinking water sources (Heaven *et al.*, 2000; Liu *et al.*, 2012; Yan *et al.*, 2009) in many countries worldwide.

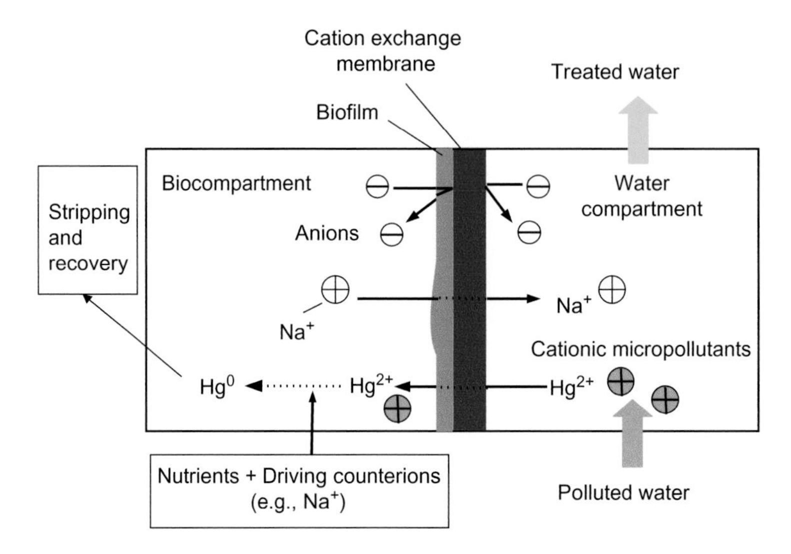

Figure 5.4. Ion-exchange membrane bioreactor (IEMB) schematic for mercury removal.

The advantages of Donnan dialysis-based processes for mercury removal from drinking water are similar to the case of As, where continuous operation without regeneration can be employed and the hydraulic retention times in both the feed and stripping compartments can be independently adjusted. Unlike the case of As, Hg(II), the most common form of Hg in water supplies, can be biologically reduced to Hg(0), which has a very low solubility in water. Therefore, mercury can be selectively removed and recovered from the water phase through the ion-exchange membrane bioreactor (IEMB) process, illustrated in Figure 5.4 (Oehmen *et al.*, 2006). The negatively charged cation-exchange membrane excludes the transport of similarly charged anions, permitting the flow of cations (e.g., Na^+) for counter transport, according to Donnan dialysis principles. The bioreduction of Hg(II) to Hg(0) in the biocompartment keeps its concentration at low levels, ensuring an adequate driving force for transport. The Hg(0) that is produced is then stripped from solution through the aeration gas and selectively recovered through sorption onto various materials, e.g., activated carbon.

The main concern of applying biological processes for drinking water treatment purposes in most cases is the risk of secondary pollution by cells and the accumulation of incompletely degraded nutrients and metabolic by-products, which can promote microbial growth in water distribution systems. The integration of bioremediation with Donnan dialysis in the IEMB prevents these undesirable attributes associated with biological processes for drinking water treatment, and provides an environmentally friendly means of removing and recovering this toxic heavy metal.

The choice of cation-exchange membrane to be employed in the IEMB process was based on a series of batch experiments operated under Donnan dialysis conditions in a stirred diffusion cell. The initial flux of Hg(II) through 11 commercially available cation-exchange membranes was assessed, and the results are shown in Figure 5.5. While some membranes, such as the Nafion, PCA and two Fumatech membranes, exhibited a substantial Hg flux to the stripping compartment, a very low rate of Hg transport was observed through the other membranes tested. A substantial portion of Hg(II) was likely retained in the membrane in these latter cases. The explanation for this result likely relates to the membrane properties, where comparatively thicker membranes (>200 μm – i.e., Ionics, Neosepta and Fumatech FTCM) require longer time to reach steady-state conditions.

The transport mechanism of Hg(II) through the membranes also depends on the speciation of Hg(II), which is pH dependent. In water at near-neutral pH, Hg(II) is largely present in the form of

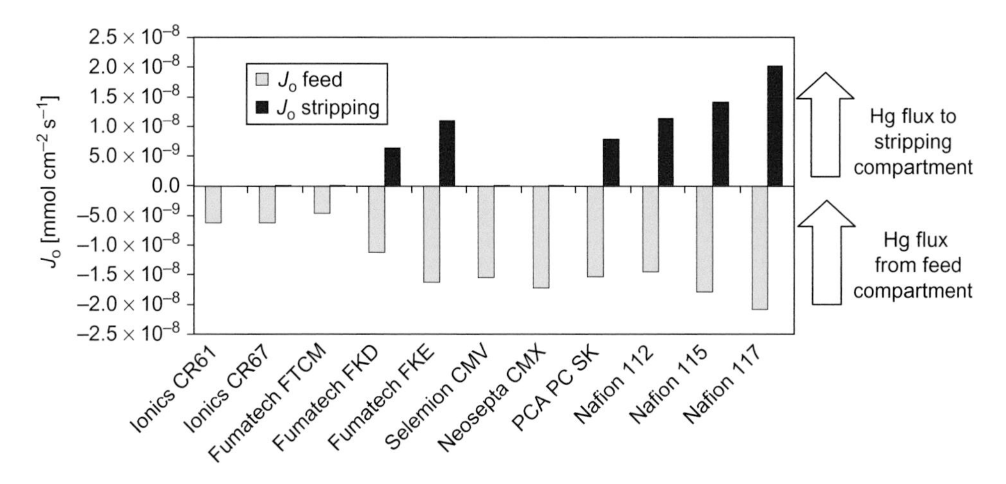

Figure 5.5. Hg(II) transport through cation-exchange membranes through Donnan dialysis (0.01 mM of $HgCl_2$ was initially present in the "feed" water compartment).

Table 5.1. Hg removal from drinking water through Donnan dialysis at a water throughput (i.e., flow rate per membrane area) of 3.1 L m^{-2} h^{-1}.

	Contaminated water	Treated water	Stripping solution
Hg [μg L^{-1}]	12.0 ± 1.6	0.6 ± 0.2	<0.2

$HgCl_2$ and $Hg(OH)_2$, where a comparatively smaller fraction exists as Hg^{2+}, $Hg(OH)^+$ and $HgCl^+$. Nevertheless, the continuous transport of the positively charged components shifts the chemical equilibrium towards the formation of additional charged compound that is then transported across the membrane. Moreover, the negatively-charged cation-exchange membranes possess a lower pH than the aqueous solutions, further accelerating the shift in chemical equilibrium of the mercuric compounds within the membrane towards the cationic form. It should also be noted that a fraction of the Hg(II) can also be transported through molecular diffusion, in addition to the Donnan counter-ion based exchange mechanism.

In order to minimize the time required to achieve steady-state flux conditions, the Fumatech FKE membrane was selected for further study via the IEMB system, considering also the comparatively low cost of this membrane as compared to e.g., Nafion membranes. Further tests were performed at lower initial Hg concentrations in order to assess the suitability of the process for achieving Hg removal from contaminated drinking water. It was found that Hg removal to levels below the 1 μg L^{-1} limit could be readily achieved (Table 5.1), although Hg was not detected in the stripping compartment, even before inoculation with Hg(II) reducing biomass.

The reason for this is likely due to the extended time required with relatively dilute solutions to equilibrate the membrane and achieve steady-state conditions. One potential alternative could be to pre-treat the membrane with a $HgCl_2$ solution in order to accelerate this process.

In cases where steady-state conditions are achieved, Donnan dialysis can be effectively integrated with Hg(II) bioreduction to Hg(0) using mixed microbial cultures. While many organisms have been shown to perform Hg(II) bioreduction in pure culture, mixed microbial cultures offer the advantage of not requiring aseptic operational conditions, and can thus be much more readily integrated with the IEMB process, where aseptic conditions would involve additional operational costs and complicate the process unnecessarily. Efficient Hg(II) bioreducing mixed cultures have been previously developed (Oehmen *et al.*, 2009) and permit the selective recovery of Hg(0) in the off-gas. This allows for simultaneous treatment of the drinking water and brine solutions, thereby minimizing the quantity of contaminated waste that is generated through the IEMB process and reducing its environmental impact.

5.3 CONCLUDING REMARKS AND FUTURE NEEDS

- As it was shown in this chapter, electromembrane processes can be successfully used for the removal of trace toxic metals from water. While laboratory bench-scale processes have been used for process development and optimization, the next steps are the validation of this process at the pilot-scale before implementation of large-scale units in the drinking water industry.
- Uranium in the form of uranyl cations, fluorine as fluoride ion and arsenic in form of arsenate can be efficiently removed by electro-membrane processes. The use of DD appears to be more indicated than that of ED, especially in the case of treating drinking water supplies, since in the latter case, high water demineralization rates, electroosmosis, and membrane scaling related issues can emerge. Moreover, although in general allowing for higher target ion fluxes than DD the elevated energy costs for ED can become an important issue in case of presence of high levels of accompanying ions in the water to be treated. This can significantly limit the ED feasibility as a process of choice. As a general rule, the lower the target toxic ion concentration in the treated water, and the higher the amounts of accompanying ions in the feed water, the less indicated becomes the application of an ED process. Special care must be also taken to possible anion-exchange competitions (higher selectivies of strong anion-exchange resin-based membranes for chloride over fluoride and for sulfate and arsenate) when the objective is fluoride or arsenate removal from water, respectively.
- Nowadays, the relatively high cost of ion-exchange membranes is one of the main limitations towards wider use of electromembrane processes in practice. Since these membranes are industrially produced almost exclusively with a flat geometry (for electrodialysis process applications), the membrane to volume ratio (compactness) of the apparatuses and the control of the hydrodynamic conditions in the chambers are limited. On the other hand, for Donnan dialysis applications, hollow fiber membrane modules seem to be more suitable since no electrodes are required for process operation. Therefore, the development of appropriate ion-exchange membranes with hollow fiber geometry would make possible the design of more compact Donnan dialysis systems, which would permit operation under more efficient water fluid dynamic conditions, by circulating the stream containing trace toxic metal(s) inside the lumen of the fibers.
- In the absence of suitable hollow fiber anion-exchange membranes, plate-and-frame configurations have to be used. The most significant aspect that has to be taken into consideration is the design of the flow channels in both the biological and water stream sides, in order to avoid clogging in the biological channel and provide adequate fluid dynamic conditions in the water channel. The design of dedicated and appropriate spacers to be introduced in the water channel, assuring reduced mass transfer resistance, low pressure drop and reduced energy input, is of major importance.
- The route towards sustainable water treatment raises new challenges to electromembrane processes. From a life cycle analysis viewpoint, sustainability has to take into account the various process stages – including the manufacturing of ion-exchange membranes, which have to respect the principles of "Green Chemistry", and the process material and energy demands, where the fate of the waste streams, such as the brine solutions, must be treated and/or disposed. The latter applies also to the "end of life" of these very membranes. The main challenges involve sustainable development and manufacturing of ion-exchange membranes with improved properties in terms of their exchange capacity, selectivity and operational stability, while reducing the environmental impact associated with their production and use in electromembrane processing. Furthermore, ion-exchange membranes, which have been used to remove toxic metals,

may slowly release them, if they are still attached to the membrane polymeric matrixes. Burning ion-exchange membranes, if not incinerated properly, may release toxic and odorous fumes.
- In what concerns the treatment/disposal of brine solutions, future approaches are expected to increasingly involve the development of integrated (hybrid) processes aiming at minimization of the volume and ecotoxicity of the brine streams, either through combining the electromembrane transport of a toxic metal with its simultaneous coagulation, adsorption, electrochemical oxidation/reduction at appropriate electrodes in a separate compartment or by volatilization (see the case of mercury removal through the IEMB concept described in this chapter).

The overall process sustainability has to be evaluated in terms of a life-cycle analysis perspective, where treatment schemes involving electromembrane processing for the removal of toxic metals from water must be benchmarked against other alternative routes.

ACKNOWLEDGEMENTS

The financial support by *Fundação para a Ciência e a Tecnologia* through projects PPCDT/AMB/57356/2004 and PEst-C/EQB/LA0006/2013 is acknowledged. A number of MSc and PhD students and other grant holders have been involved in the technical and/or analytical work related to development and validation of the IEMB (IEMR) concepts. The authors gratefully acknowledge their tireless dedication and great enthusiasm.

REFERENCES

Abou-Shady, A., Peng, C., Bi, J., Xu, H. & Almeria, J. (2012) Recovery of Pb (II) and removal of NO_3^- from aqueous solutions using integrated electrodialysis, electrolysis, and adsorption process. *Desalination*, 286, 304–315.

Amor, Z., Bariou, B., Mameri, N., Taky, M., Nicolas, S. & Elmidaoui, A. (2001) Fluoride removal from brackish water by electrodialysis. *Desalination*, 133, 215–223.

Banasiak, L.J. (2009) *Removal of inorganic and trace organic contaminants by electrodialysis*. PhD Thesis. The University of Edinburgh, Edinburgh, UK.

Barringer, J.L. & Szabo, Z. (2006) Overview of investigations into mercury in ground water, soils, and septage, New Jersey coastal plain. *Water, Air, & Soil Pollution*, 175, 193–221.

Bergman, R.A. (1995) Membrane softening *versus* lime softening in Florida: a cost comparison update. *Desalination*, 102, 11–24.

Crespo, J.G. & Reis, M.A.M. (2001) Treatment of aqueous media containing electrically charged compounds. Patent WO 0,140,118.

Davis, T. (2000) Donnan dialysis. In: Wilson, I.D. (ed.) *Encyclopedia of separation science*. Volume 4. Academic Press, London, UK. pp. 1701–1707.

Davis, T.A., Wu, J.S. & Baker, B.L. (1971) Use of the Donnan equilibrium principle to concentrate uranyl ions by an ion-exchange membrane process. *AIChE Journal*, 17 (4), 1006–1008.

DeZuane, J. (1997) *Handbook of drinking water quality*. 2nd edition. John Wiley & Sons, New York, NY.

Dieye, A., Larchet, C., Auclair, B. & Mar-Diop, C. (1998) Elimination des fluorures par la dialyse ionique croisée. *European Polymer Journal*, 34, 67–75.

Dzyazko, Y.S. & Belyakov, V.N. (2004) Purification of a diluted nickel solution containing nickel by a process combining ion exchange and electrodialysis. *Desalination*, 162, 179–189.

Feng, X., Wu, Z. & Chen, X. (2007) Removal of metal ions from electroplating effluent by EDI process and recycle of purified water. *Separation and Purification Technology*, 57, 257–263.

Garcia, R. (2006) *Cerca de 60 mil portugueses bebem água com arsénio a mais*. Público Lisboa, Portugal, pp. 1–2 (in Portuguese).

Garmes, H., Persin, F., Sandeaux, J., Pourcelly, G. & Mountadar, M. (2002) Defluoridation of groundwater by a hybrid process combining adsorption and Donnan dialysis. *Desalination*, 145, 287–291.

Grebenyuk, V.D., Chebotareva, R.D., Linkov, N.A. & Linkov, V.M. (1998) Electromembrane extraction of Zn from Na-containing solutions using hybrid electrodialysis-ion exchange method. *Desalination*, 115, 255–263.

Grimm, J., Bessarabov, D. & Sanderson, R. (1998) Review of electro-assisted methods for water purification. *Desalination*, 115, 285–294.

Guell, R., Fontàs, C., Anticó, E., Salvadó, V., Crespo, J.G. & Velizarov, S. (2011) Transport and separation of arsenate and arsenite from aqueous media by supported liquid and anion-exchange membranes. *Separation and Purification Technology*, 80, 428–434.

Heaven, S., Ilyushchenko, M.A., Tanton, T.W., Ullrich, S.M. & Yanin, E.P. (2000) Mercury in the river Nura and its floodplain. Central Kazakhstan. I. River sediments and water. *Science of the Total Environment*, 260, 35–44.

Henke, K.R. (2009) *Arsenic: environmental chemistry, health threats, and waste treatment*. 1st edition. John Wiley & Sons, New York, NY.

Hichour, M., Persin, F., Sandeaux, J. & Gavach, C. (2000) Fluoride removal from waters by Donnan dialysis. *Separation and Purification Technology*, 18, 1–11.

Jacangelo, J.G., Trussell, R.R. & Watson, M. (1997) Role of membrane technology in drinking water treatment in the United States. *Desalination*, 113, 119–127.

Jiang, G.-B., Shi, J.-B. & Feng, X.-B. (2006) Mercury pollution in China. An overview of the past and current sources of the toxic metal. *Environmental Science & Technology*, 40, 3672–3678.

Kabay, N., Arar, O., Samatya, S., Yüksel, U. & Yüksel, M. (2008) Separation of fluoride from aqueous solution by electrodialysis: effect of process parameters and other ionic species. *Journal of Hazardous Materials*, 153, 107–113.

Koter, S. & Warszawski, A. (2000) Electromembrane processes in environmental protection. *Polish Journal of Environmental Studies*, 9, 45–56.

Lisha, K., Anshup, P. & Pradeep, T. (2009) Towards a practical solution for removing inorganic mercury from drinking water using gold nanoparticles. *Gold Bulletin*, 42, 144–152.

Liu, B., Yan, H., Wang, C., Li, Q., Guedron, S., Spangenberg, J.E., Feng, X. & Dominik, J. (2012) Insights into low fish mercury bioaccumulation in a mercury-contaminated reservoir, Guizhou, China. *Environmental Pollution*, 160, 109–117.

Mahmoud, A. & Hoadley, A.F. (2012) An evaluation of a hybrid ion exchange electrodialysis process in the recovery of heavy metals from simulated dilute industrial wastewater. *Water Research*, 46, 3364–3376.

Matos, C.T., Velizarov, S., Reis, M.A. & Crespo, J.G. (2008) Removal of bromate from drinking water using the ion exchange membrane bioreactor concept. *Environmental Science & Technology*, 42, 7702–7708.

Montana, M., Camacho, A., Serrano, I., Devesa, R., Matia, L. & Valles, I. (2013) Removal of radionuclides in drinking water by membrane treatment using ultrafiltration, reverse osmosis and electrodialysis reversal. *Journal of Environmental Radioactivity*, 125, 86–92.

Monzie, I., Muhr, L., Lapicque, F. & Grévillot, G. (2005) Mass transfer investigations in electrodeionization processes using the microcolumn technique. *Chemical Engineering Science*, 60, 1389–1399.

Nataraj, S.K., Sridhar, S., Shaikha, I.N., Reddy, D.S. & Aminabhavi, T.M. (2007) Membrane-based microfiltration/electrodialysis hybrid process for the treatment of paper industry wastewater. *Separation and Purification Technology*, 57, 185–192.

Nies, D.H. (1999) Microbial heavy-metal resistance. *Applied Microbiology and Biotechnology*, 51, 730–750.

Oehmen, A., Viegas, R., Velizarov, S., Reis, M.A.M. & Crespo, J.G. (2006) Removal of heavy metals from drinking water supplies through the ion exchange membrane bioreactor. *Desalination*, 199, 405–407.

Oehmen, A., Fradinho, J., Serra, S., Carvalho, G., Capelo, J.L., Velizarov, S., Crespo, J.G. & Reis, M.A.M. (2009) The effect of carbon source on the biological reduction of ionic mercury. *Journal of Hazardous Materials*, 165, 1040–1048.

Oehmen, A., Valerio, R., Llanos, J., Fradinho, J., Serra, S., Reis, M.A.M., Crespo, J.G. & Velizarov, S. (2011) Arsenic removal from drinking water through a hybrid ion exchange membrane – coagulation process. *Separation and Purification Technology*, 83, 137–143.

Ricardo, A.R., Carvalho, G., Velizarov, S., Crespo, J.G. & Reis, A.M. (2012) Kinetics of nitrate and perchlorate removal and biofilm stratification in an ion exchange membrane bioreactor. *Water Research*, 46, 4556–4568.

Richardson, S.D. (2003) Disinfection by-products and other emerging contaminants in drinking water. *TrAC Trends in Analytical Chemistry*, 22, 666–684.

Sahli, M.A., Annouar, S., Tahaikt, M., Mountadar, M., Soufianec, A. & Elmidaoui, A. (2007) Fluoride removal for underground brackish water by adsorption on the natural chitosan and by electrodialysis. *Desalination*, 212, 37–45.

Shih, M.C. (2005) An overview of arsenic removal by pressure-driven membrane processes. *Desalination*, 172, 85–97.

Smith, A.H., Lopipero, P.A., Bates, M.N. & Steinmaus, C.M. (2002) Arsenic epidemiology and drinking water standards. *Science*, 296, 2145–2146.
Spoor, P.B., Koene, L., Ter Veen, W.R. & Janssen, L.J.J. (2002a) Continuous deionization of a dilute nickel solution. *Chemical Engineering Journal*, 85, 127–135.
Spoor, P.B., Koene, L. & Janssen, L.J.J. (2002b) Potential and concentration gradients in a hybrid ion-exchange/electrodialysis cell. *Journal of Applied Electrochemistry*, 32, 369–377.
Strathmann, H. (2004) Ion-exchange membrane separation processes. In: Burggraaf, A.J. & Cot, L. (eds.) *Membrane science and technology series*. Volume 9. Elsevier, Amsterdam, The Netherlands. pp. 227–286.
Strathmann, H. (2010) Electrodialysis, a mature technology with a multitude of new applications. *Desalination*, 264, 268–288.
Tran, A.T.K., Zhang, Y., Jullok, N., Meesschaert, B., Pinoy, L. & Van der Bruggen, B. (2012) RO concentrate treatment by a hybrid system consisting of a pellet reactor and electrodialysis. *Chemical Engineering Science*, 79, 228–238.
USEPA (2001) Panel 14: National primary drinking water regulations: arsenic and clarifications to compliance and new source contaminants monitoring. Volume 66 (194). Environmental Protection Agency, Washington, DC.
Van der Bruggen, B., Manttari, M. & Nystrom, M. (2008) Drawbacks of applying nanofiltration and how to avoid them: a review. *Separation and Purification Technology*, 63, 251–263.
Velizarov, S. (2003) Removal of trace mono-valent inorganic pollutants in an ion exchange membrane bioreactor: analysis of transport rate in a denitrification process. *Journal of Membrane Science*, 217, 269–284.
Velizarov, S., Reis, M.A. & Crespo, J.G. (2013) Transport of arsenate through anion-exchange membranes in Donnan dialysis. *Journal of Membrane Science*, 425–426, 243–250.
Velizarov, S., Crespo, J.G. & Reis, M.A. (2004) Removal of inorganic anions from drinking water supplies by membrane bio/processes. *Reviews in Environmental Science and Bio/Technology*, 3, 361–380.
Velizarov, S., Matos, C., Reis, M. & Crespo, J. (2005) Removal of inorganic charged micropollutants in an ion-exchange membrane bioreactor. *Desalination*, 178, 203–210.
Velizarov, S., Reis, M.A. & Crespo, J.G. (2011) The ion-exchange membrane bioreactor: developments and perspectives in drinking water treatment. Chapter 4 in: Coca-Prados, J. & Gutiérrez-Cervelló, G. (eds.) Water purification and management. *NATO Science for Peace and Security Series* C: *Environmental Security*. Springer Science + Business Media B.V. pp. 119–145.
Wallace, R.M. (1967) Concentration and separation of ions by Donnan membrane equilibrium. *Industrial & Engineering Chemistry Process Design and Development*, 6 (4), 423–431.
WHO (2008) Guidelines for drinking-water quality. Volume 1: Recommendations. 3rd edition. World Health Organisation, Geneva, Switzerland.
Yan, H., Feng, X., Shang, L., Qiu, G., Dai, Q., Wang, S. & Hou, Y. (2008) The variations of mercury in sediment profiles from a historically mercury-contaminated reservoir. Guizhou province, China. *Science of the Total Environment*, 407, 497–506.
Zaheri, A., Moheb, A., Keshtkar, A.R. & Shirani, A.S. (2010) Uranium separation from wastewater by electrodialysis. *Iranian Journal of Environmental Health Science & Engineering*, 7, 429–436.
Zaki, E.E. (2002) Electrodialysis of uranium(VI) through cation exchange membranes and modeling of electrodialysis processes. *Journal of Radioanalytical and Nuclear Chemistry*, 252, 21–30.
Zhao, B., Zhao, H. & Ni, J. (2010) Modeling of the Donnan dialysis process for arsenate removal. *Chemical Engineering Journal*, 160, 170–175.
Zhao, B., Zhao, H., Dockko, S. & Ni, J. (2012) Arsenate removal from simulated groundwater with a Donnan dialyzer. *Journal of Hazardous Materials*, 215–216, 159–165.

CHAPTER 6

Fluoride, arsenic and uranium removal from water using adsorbent materials and integrated membrane systems

Hacene Mahmoudi, Noreddine Ghaffour & Mattheus Goosen

6.1 INTRODUCTION

Depending on the region and the geology, untreated water may contain, naturally occurring toxic elements, such as arsenic (As), uranium (U) and fluoride (F^-). These elements, which are considered as extremely poisonous, are directly transmitted to people when the untreated water is used for drinking, food preparation, recreation, or for various domestic purposes (Bhatnagar *et al.*, 2011; Iakovleva and Sillanpää, 2013; Nordstrom, 2002). As a result millions of people around the world are threatened by F^-, As and U contamination. There is, thus, a need for low-cost and proven technologies that can effectively treat polluted water especially in developing countries. In this chapter, novel and conventional techniques are critically reviewed for the removal of these toxic contaminants from groundwater and wastewater.

6.2 FLUORIDE

6.2.1 *Introduction*

A comprehensive review on defluoridation of drinking water with an emphasis on the use of sustainable technologies was written by Ayoob *et al.* (2008). The authors critically compared different processes. It was concluded by them that while traditional coagulation methods have generally been found to be effective in defluoridation, they were unsuccessful in reducing F^- concentrations to desired low levels. Furthermore, they maintained that while newer technologies such as membrane processes do not require additives, the technology is relatively expensive to install and operate and the membranes are prone to fouling, scaling, or degradation.

6.2.2 *An overview of technologies for fluoride removal from water*

In a recent review containing over 200 references, Bhatnagar *et al.* (2011) reported on traditional adsorption methods for removing F^- from drinking water including liming (i.e., addition of calcium hydroxide) and the accompanying precipitation of fluorite, the precipitation and coagulation processes with iron (III), activated alumina, alum sludge and calcium, and ion exchange (Table 6.1). The authors argued that shortcomings of most of these methods included high operating and upkeep costs, secondary contamination such as production of a toxic sludge by-product, and complex treatment processes.

Ayoob *et al.* (2008) highlighted the basic principles and procedures involved in current F^- elimination technologies. They reported that defluoridation techniques can be generally grouped into coagulation, adsorption and/or ion exchange, electrochemical, and membrane processes. The coagulation technique involves precipitation or coprecipitation of F^- by using suitable reagents like lime, calcium and magnesium salts, polyaluminum chloride, and alum. Adsorption is another important technique most widely used for excess F^- removal from aqueous solution. In this process a packed bed of adsorbent in fixed columns is continuously used for cyclic sorption and/or desorption of pollutants by effectively utilizing the capacity of the bed. The adsorbents

Table 6.1. Adsorption techniques for fluoride removal from water (adapted from Bhatnagar *et al.*, 2011).

Adsorbent	Examples
Alumina & aluminum based adsorbents	Fluoride (F^-) binds to $Al(OH)_3$ and Al_2O_3
Calcium-based sorbents	Affinity of calcium (Ca^{2+}) for fluoride anion
Iron-based sorbents	Granular ferric hydroxide $Fe(OH)^3$ binds F^-
Metal oxides/hydroxides	Cerium-based adsorbent
Carbon-based adsorbents	Small grain sizes for better F^- adsorption
Natural materials as sorbents	Fluoride binds to bituminous coal & clay
Biosorbents	Chitosan with NH_3^+ & OH^- binds fluoride
Agricultural wastes	Corn cob powder for fluoride adsorption
Industrial wastes as sorbents	Carbon slurry from fertilizer industry
Hydroxyapatite	Highly porous for large adsorption surface
Nano-sorbents	Carbon nanotubes with large surface area

generally used include bone char, activated alumina, activated carbon, activated bauxite, ion-exchange resins, fly ash, super phosphate and tricalcium phosphate, clays and soils, synthetic zeolites, and other minerals.

Electrochemical techniques include electrocoagulation and involve the use of aluminum electrodes that release Al^{3+} ions by an anodic reaction, and the ions then react with F^- ions that are found in excess near the anode. Here, precipitation and thereby removal of F^- occurs at the electrode/electrolyte interface. In addition, membrane techniques include reverse osmosis, nano-/ultrafiltration, and electrodialysis. Defluoridation based on a combination of two or more of these processes have also been reported (Hu *et al.*, 2003; Mjengera and Mkongo, 2003; Velizarov *et al.*, 2004). Mjengera and Mkongo (2003) for example determined that the bone char method was appropriate for use in rural areas of Tanzania suffering from excessive F^- in their water sources due to its simplicity, local availability of materials and the possibility of processing the material locally (Fig. 6.1a). The authors also reported on an institution level defluoridation plant based on the alum and lime method (Fig. 6.1b). While effective it can be argued that such a complex system would not be feasible for poor rural areas.

6.2.3 *Fluoride removal from drinking water by adsorption on naturally occurring biopolymers*

Kamble *et al.* (2007) reported on the applicability of chitin, chitosan and chemically modified chitosan (20%-lanthanum chitosan) as adsorbents for the removal of excess F^- from drinking water. Chitosan which is derived from chitin is one of the main components of crustacean shells of prawn, crab, shrimp or lobster, has the ability to coordinate metal ions because of its high concentration of amine functional groups (Li *et al.*, 1992). It is also a non-toxic, biodegradable and biocompatible material. Furthermore, the effects of various physico-chemical parameters such as pH, adsorbent dose, initial F^- concentration and the presence of interfering ions on adsorption of F^- were assessed by Kamble *et al.* (2007). The authors concluded that lanthanum chitosan adsorbents were better at removal of F^- from water than plain chitosan and chitin (Fig. 6.2).

The adsorption of F^- on the surface of the adsorbent was found to depend mainly on the pH of the solution as well as the concentration and type of co-anions (Fig. 6.3). Kamble *et al.* (2007) also established that the presence of anions has a deleterious effect on the adsorption of F^-, particularly carbonate and bicarbonate anions. The mechanism of adsorption of F^- on lanthanum (La) modified chitosan was explained in terms of the ligand exchange mechanism between F^- ion and hydroxide ion coordinated to La(III) ion immobilized on the chitosan.

The percentage removal of F^- in distilled water was also observed to be higher than ground water; this may have been due to the fact that the latter contains different types of ions. No significant leaching of lanthanum was observed from the adsorbent. It was also possible to regenerate the material, which is important for sustainability.

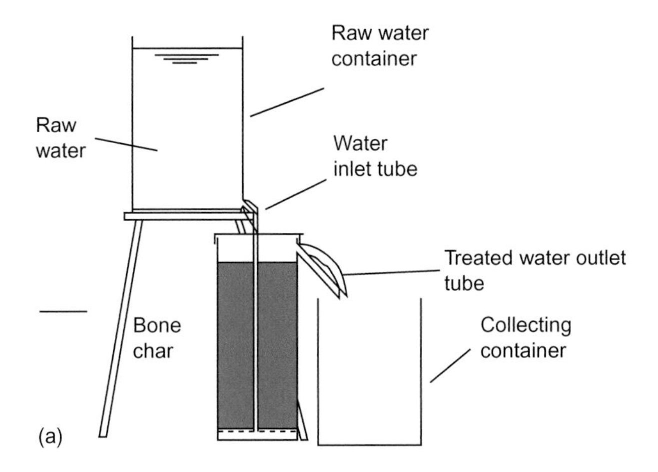

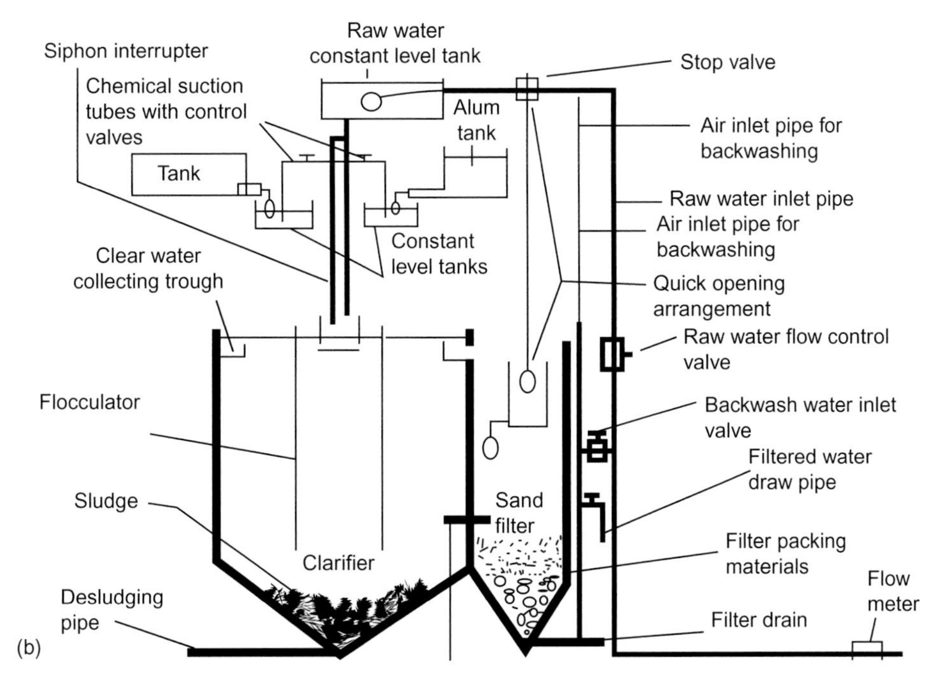

Figure 6.1. (a) Arrangement of household bone char filter column defluoridation unit; (b) an institution level defluoridation plant based on alum and lime method (adapted from Mjengera and Mkongo, 2003).

In a related study, a biomass material, bone char was investigated by Ma *et al.* (2008) for its feasibility as a cost-effective biosorbent for F^- removal from drinking water. The amorphous biosorbent powder, which is composed mainly of calcium phosphate and a small amount of carbon, was prepared by heating bone-biomass. The adsorption capacity of the bone char was shown to be better than that of activated aluminum and tourmaline (i.e., crystal boron silicate mineral compounded with elements such as aluminum, iron, magnesium, sodium, lithium, or potassium). Removal of F^- was attributed to the processes of ion binding and ion exchange between bone char and F^-. The authors developed static and kinetic models which provided a satisfactory prediction of F^- concentration after adsorption. Experiments with fixed-bed columns

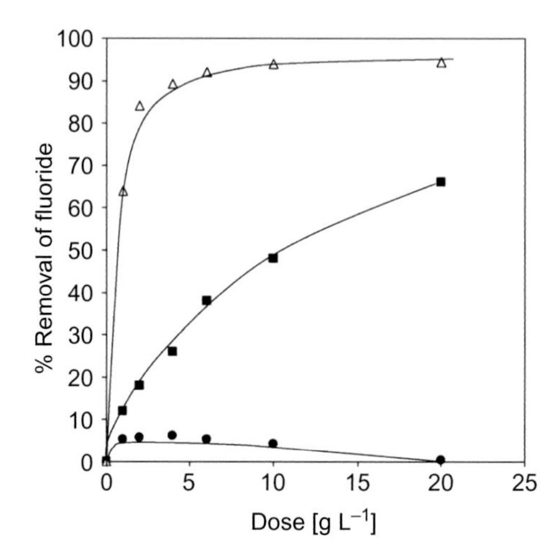

Figure 6.2. Comparison of chitin, chitosan and 20% La-chitosan for fluoride removal (pH 6.7, contact time = 24 h). (•) Chitin; (■) chitosan; (△) 20% La-chitosan (Kamble *et al.*, 2007).

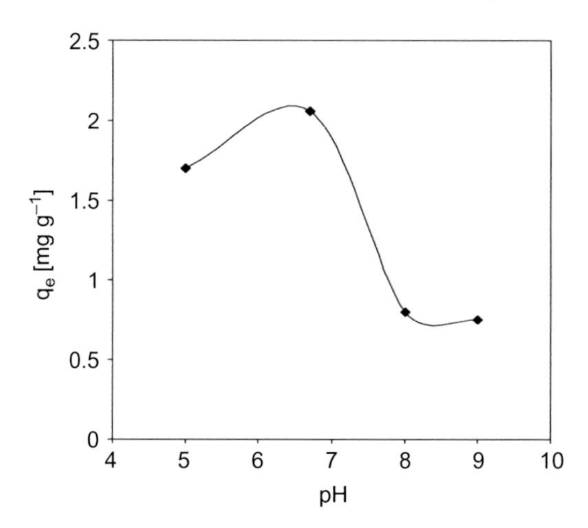

Figure 6.3. Effect of pH on adsorption of fluoride (initial concentration = 5 mg L^{-1}; optimum dose = 2 g L^{-1}; contact time = 24 h) (Kamble *et al.*, 2007).

indicated that adsorption capacity depends strongly on the water flow rate, inlet F^- concentration, and adsorbent column height. In addition, Ma *et al.* (2008) were able to regenerate the bone char powder using 0.5% NaOH, making it a promising material for sustainable purification of drinking water. Compared with traditional F^- removal methods, the authors argued that bone char can be used as a cost-effective biosorbent for efficient F^- removal from groundwater.

6.2.4 *Aluminum and iron oxides and bauxite as adsorbents for fluoride removal from water*

The fluoride ion, F^-, has a strong affinity for metal ions such as Al^{3+} and Fe^{3+} (Wu *et al.*, 2007). Scattering a combination of these metals in a permeable material would afford a high F^- adsorption capacity (Tchomgui-Kamga *et al.*, 2010). As an example of this type of approach, Chen *et al.* (2011) effectively developed an adsorbent by impregnation of porous granular ceramics with aluminum and iron salts to remove F^- from aqueous solution. The Al/Fe dispersed in porous granular ceramic adsorbent was reddish brown and 2–3 mm in diameter (Fig. 6.4b).

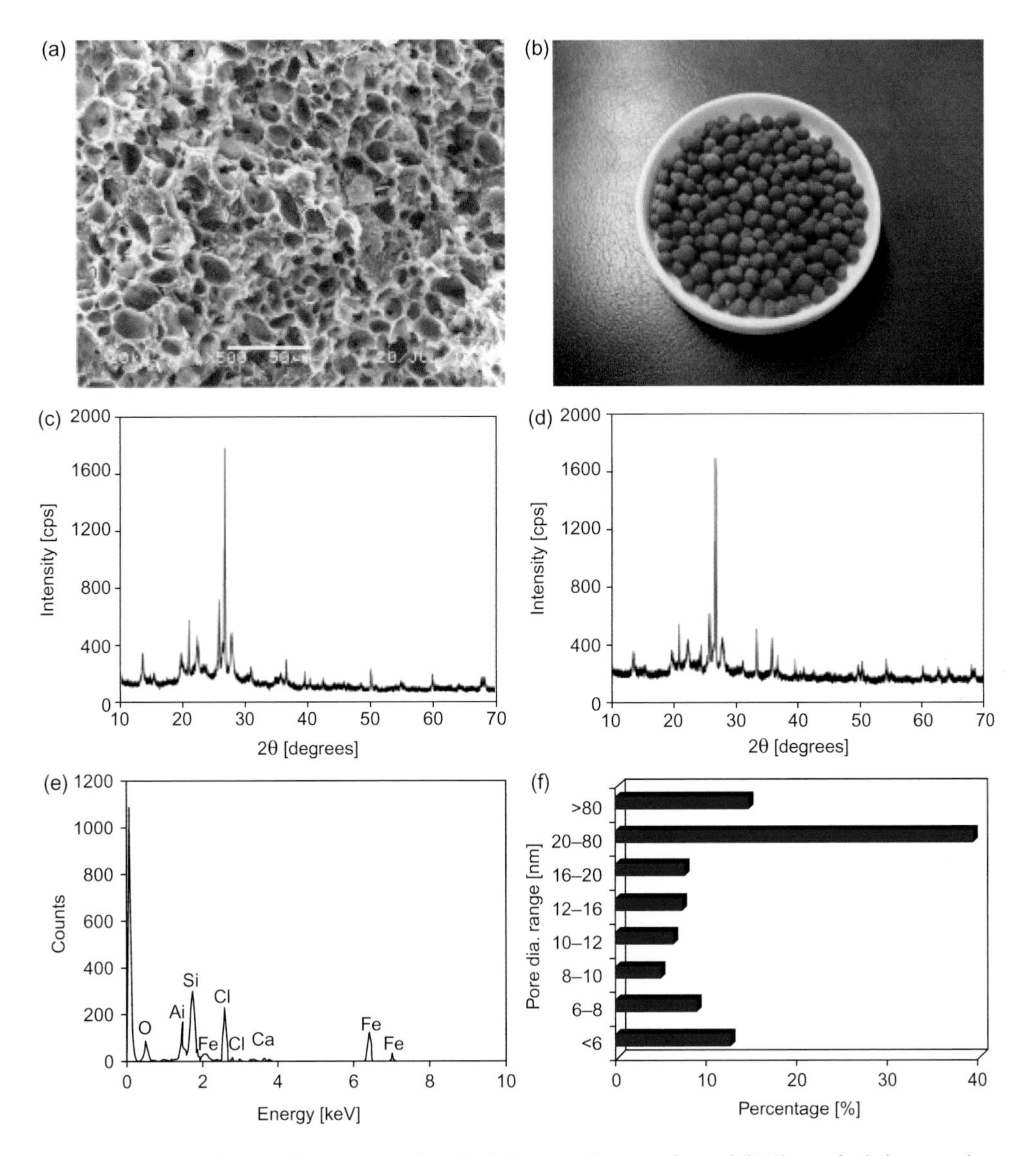

Figure 6.4. SEM images of (a) cross section of pristine granular ceramics and (b) Photo of pristine granular ceramics, Powder XRD patterns of (c) pristine granular ceramics and (d) adsorbed granular ceramics, EDS spectra of (e) pristine granular ceramics, BJH (Barrett–Joyner–Halenda) pore-size distribution of (f) pristine granular ceramics (Chen *et al.*, 2011).

As can be seen from Figure 6.4a, the adsorbent cross section had a very porous structure implying a high adsorption capacity. This pore texture was attributed to the sintering process (i.e., making the granules from powder). The EDS spectrum of Figure 6.4e showed the presence of Fe, Al, Si, O and Cl in the surface of adsorbent, which was attributed to the impregnation process with $AlCl_3$ and $FeCl_3$ salt solutions. Energy-dispersive X-ray spectroscopy (EDS) is an analytical technique normally used for the elemental analysis or chemical characterization of a sample.

Chen *et al.* (2011) argued that these low cost adsorbents showed a good efficiency in F^- removal from aqueous solution and could be useful for environmental protection purposes. The loading capacity of these porous granular ceramics with aluminum and iron salts for F^- was found to be

1.79 mg g^{-1}. The optimum level of F^- removal was observed at pH ranges of 4–9. The presence of carbonate and phosphate ions had a highly negative effect on the F^- removal capacity. Kinetic studies indicated that the adsorption process followed a pseudo-second-order kinetic model. The authors concluded that porous granular ceramics with mixed aluminum and iron oxides have a great potential for F^- removal from ground and drinking water.

In an associated study, Sujana and Anand (2011) confirmed the adsorption efficiency of bauxite for F^- removal from synthetic as well as ground water samples. The adsorption of F^- was highly dependent on pH, temperature and initial adsorbate concentrations in the solutions.

The optimum pH range for F^- adsorption on the bauxite surface was found to be 5 to 7, which was in a similar though slightly narrower range than that observed by Chen *et al.* (2011). Furthermore, a kinetic study by Sujana and Anand (2011) revealed that F^- adsorption on the bauxite surface followed first order with the Langmuir adsorption capacity being 6.16 mg g^{-1}, which was about three times as high as that found for porous granular ceramics with mixed aluminum and iron oxides (Chen *et al.*, 2011). In Sujana and Anand's (2011) work the presence of competing anions like sulfate, nitrate and phosphate showed an adverse effect, whereas carbonate ions only mildly affected the F^- adsorption. Since bauxite is an abundantly available mineral in many parts of the world, the authors concluded that it can provide a simple, effective and yet low cost method for removing F^- from contaminated water.

6.2.5 *Fluoride removal from industrial wastewater using electrocoagulation*

Fluoride ions were removed electrochemically from industrial wastewater by Shen *et al.* (2003) using a combined electrocoagulation and electroflotation process. The experimental results showed that weakly acidic conditions were favorable for this type of treatment, while too high or too low pH could affect the formation of the $Al(OH)_3$ flocs. The optimal retention time in their case was 20 min. Not surprisingly, cations and anions affected the removal process; Ca^{2+} for example was helpful in precipitating F^- and reducing the residual F^- concentration. In general, anions had a negative effect on F^- removal. As the authors explained this may have been due to the competitive adsorption between F^- ion and other anions. As Hu *et al.* (2003) noted, in the electrocoagulation process the F^- ions are attracted to the anode by the electric force. In this process the F^- concentration near the anode is higher than in the bulk solution. However, most of the F^- ions attracted to the anode are replaced by other anions, if F^- is not the dominant anion in solution. Therefore, the defluoridation efficiency may decline because of the presence of co-existing anions.

Khatibikamal *et al.* (2010) in a similar study also employed electrocoagulation (EC) with aluminum electrodes for removing F^- from treated industrial wastewater originating from the steel industry. Effects of different operating conditions such as temperature, pH, voltage, hydraulic retention time (HRT) and number of aluminum plates between anode and cathode plates on removal efficiency were assessed. Experimental results showed that by increasing HRT, removal efficiency increased but after 5 min changes were negligible. Therefore, the total HRT required was only 5 min. After treatment, the F^- concentration was reduced from an initial 4.0–6.0 mg L^{-1} to lower than 0.5 mg L^{-1}. The pH of the influent was established as a very important variable which affected F^- removal. The optimal pH range for the feedwater was 6.0–7.0 at which not only effective defluoridation could be achieved, but also no pH readjustment was needed after treatment. Additionally, increasing the number of aluminum plates between anode and cathode plates in the system did not significantly affect F^- removal.

6.2.6 *Elimination of fluoride from drinking and wastewater using a combination of traditional and membrane techniques*

Reverse osmosis (RO) and nanofiltration (NF) may be used to reduce the concentration of F^- in wastewater. For example Dolar *et al.* (2011) investigated the removal efficiency of RO and NF membranes to reduce fluoride and phosphate load in wastewater from fertilizer factories to

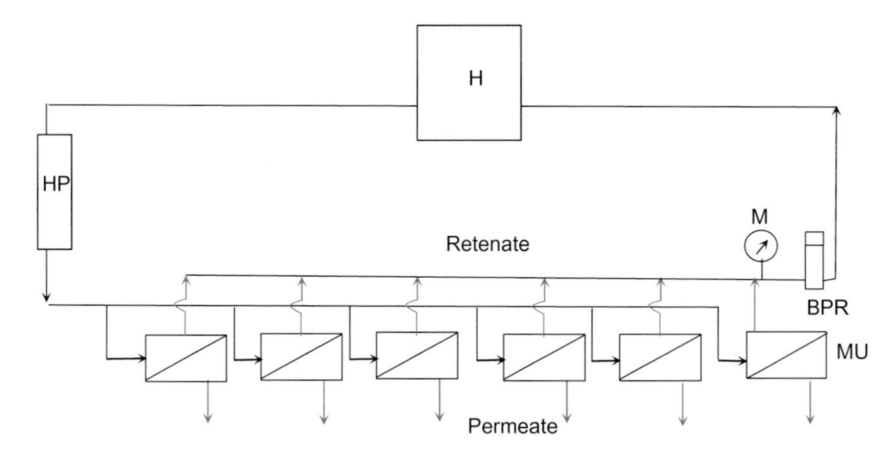

Figure 6.5. Schematic representation of reverse osmosis/nanofiltration (RO/NF) laboratory set up: H – holdup tank, HP – high pressure pump, M – manometer, MU – RO/NF cells, BPR – back pressure regulator. (Dolar *et al.*, 2011).

less than $8\,\mathrm{mg\,L^{-1}}$ and $2\,\mathrm{mg\,L^{-1}}$, respectively (Fig. 6.5). Their laboratory study indicated that the rejection of F^- with RO membranes was higher than 80% (model waters) and higher than 96% (real wastewater), and with NF membranes it was greater than 40%. However we can argue that this is not sufficient for drinking water quality. As mentioned earlier according to WHO standards, the optimum F^- level in drinking water is considered to be between 0.5 and $1.0\,\mathrm{mg\,L^{-1}}$ (Ghorai and Pantk, 2005; Wang and Reardon, 2001; WHO, 2011). In addition USEPA recently established the effluent discharge standard of $4\,\mathrm{mg\,L^{-1}}$ for F^- from wastewater treatment plant (Khatibikamal *et al.*, 2010; Shen *et al.*, 2003). This suggests that membrane techniques should be combined with traditional methods to reduce the F^- levels to WHO standards.

Kowalchuck (2011) selectively removed F^- from drinking water based on a process which combined precipitation by aluminum hydroxide $Al(OH)_3$ with subsequent removal of the floc by membrane ultrafiltration. A $0.3\,\mathrm{gal\,min^{-1}}$ ($\sim 11.4\,\mathrm{L\,min^{-1}}$) pilot test plant achieved F^- removal to a concentration of $3.5\,\mathrm{mg\,L^{-1}}$ at an aluminum dose of $30\,\mathrm{mg\,L^{-1}}$. The former F^- concentration met the USEPA effluent discharge standard of $4\,\mathrm{mg\,L^{-1}}$ for F^- from wastewater. However, it was not in the optimum range since the maximum WHO safe F^- level in drinking water is considered to be between 0.5 and $1.0\,\mathrm{mg\,L^{-1}}$ (Ghorai and Pantk, 2005; Wang and Reardon, 2001; WHO, 2011).

Removal of F^- from drinking water by a membrane coagulation reactor was assessed by Zhang *et al.* (2005), using aluminum sulfate as the major floc forming chemical. The optimum pH value for removal of F^- was found to be in the range from 6.0 to 6.7. In laboratory-scale tests, when proper dosages of aluminum sulfate and sodium hydroxide were added to the reactor, the concentration of F^- was reduced from $4.0\,\mathrm{mg\,L^{-1}}$ in raw water to less than $1.0\,\mathrm{mg\,L^{-1}}$ in the product water, which is within both the WHO and USEPA standards. These were comparable results to those observed by Kowalchuck (2011).

Membrane fouling is a serious problem in any separation process (Goosen *et al.*, 2011). Membrane properties may also be affected by the feed water temperature. Fouling of the membrane surface requires for cleaning using both physical and chemical methods (Al Obeidani *et al.*, 2008). The primary feature of a chemical cleaning process, for example, is a heterogeneous reaction between the detergent solution and the fouled layer (Tran-Ha and Wiley, 1989). The cleaning reaction can be divided into six stages (Fig. 6.6). All six stages do not necessarily always occur.

When cleaning, for example, equipment with fatty fouling, it may be necessary to melt the fat (thereafter the oil could simply be eroded away by hot water), or detergent may be added to emulsify the oil, or it may be dissolved using a suitable solvent (Luss, 1990). The chemicals used as cleaning agents should loosen and dissolve the foulant, keep the foulant in dispersion and solution, avoid spacer fouling, not attack the membrane (and other parts of the system),

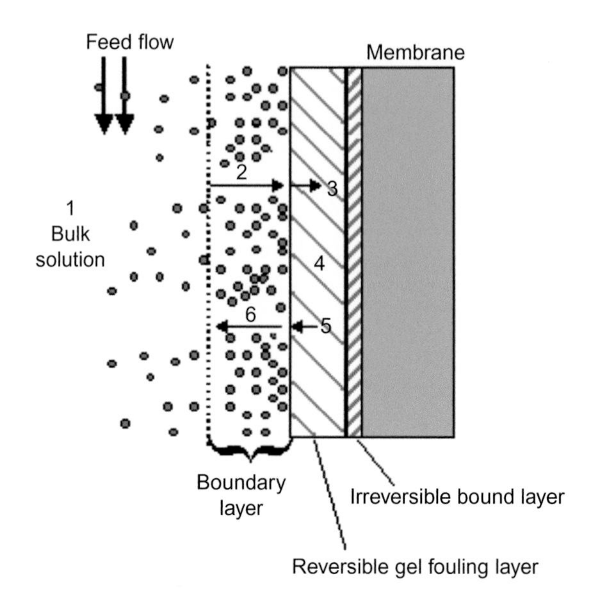

Figure 6.6. Chemical cleaning process at membrane surface: (1) Bulk reactions; (2) Transport of detergent to interface; (3) Transport of detergent into foulant layer; (4) Cleaning reactions in fouling layer; (5) Transport of cleaning reaction products back to interface; and (6) Transport of product to bulk solution (Al Obeidani *et al.*, 2008).

and disinfect all wetted surfaces (Tragardh, 1989). Besides the cleaning ability of a detergent, there are other important factors such as the ease with which it can be dispensed and rinsed away, its chemical stability during use, and cost and safety. All of these will play a key role in determining the large-scale commercial feasibility membrane techniques for removal of F^- and similar pollutants for wastewater.

Zhang *et al.* (2005) in an attempt to overcome fouling problems in the membrane coagulation reactor employed both physical and chemical cleaning to try and regenerate the original flux. While both were effective, chemical cleaning generally achieved better results. For example, at the end of run 4 physical cleaning gave a flux of $38.7\,L\,m^{-2}\,h^{-1}$ while chemical cleaning gave a flux of $46.2\,L\,m^{-2}\,h^{-1}$; the latter compares more favorably to the original flux of $49.3\,L\,m^{-2}\,h^{-1}$ for the unused membrane. In a related study, Al Obeidani *et al.* (2008) in a set of experiments used two chemical agents in a cleaning process starting with alkaline and followed by acid (Fig. 6.7). The results showed that the flux recovery was 93.5% and the operating cycle time was 85% (528 h) of the original cycle time of a new membrane (624 h).

Velizarov *et al.* (2004) in a recommended paper provided an overview of the main membrane-assisted processes that can be used for the removal of toxic inorganic anions from drinking water supplies. The authors emphasized integrated process solutions, which combined traditional and membrane techniques, including membrane bioreactors. It was concluded, for example, that merging the advantages of membrane separation with biological reactions for the treatment of polluted water supplies has resulted in the development of three major membrane bioreactors: pressure-driven, gas transfer, and ion exchange. In the first type, membranes are essentially regarded as porous barriers to promote high biomass for process amplification and avoid contamination of the treated water with microbial cells. Nevertheless, secondary water pollution by an incompletely degraded organic carbon source and other low molecular mass compounds is possible. It was reported by Velizarov *et al.* (2004) that hydrogen gas-transfer membrane bioreactors appear especially attractive for *in-situ* water remediation, while ion exchange membrane bioreactors can provide a highly selective target ion removal and avoid secondary pollution of the treated water.

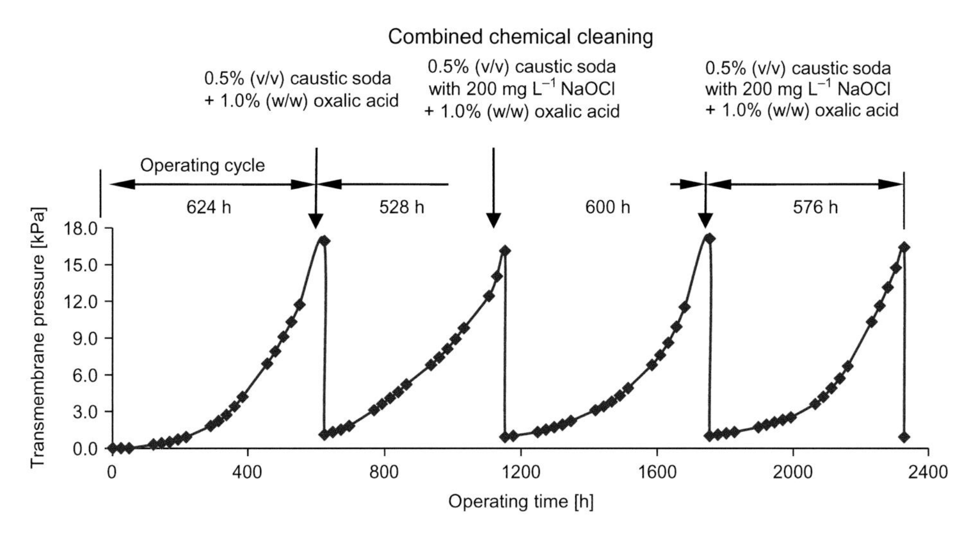

Figure 6.7. Effect of combination of both cleaning agents on flux recovery and operating time (Al Obeidani *et al.*, 2008).

6.2.7 *Potential of adsorbents/fillers in membranes (mixed-matrix membranes) for fluoride removal*

Mixed matrix membranes, a relatively recent development, take advantage of both the relatively low cost of fabricating polymeric membranes and the mechanical strength and functional properties of inorganic materials, such as zeolite, silver, silica as well as nanoparticles (Hoek *et al.*, 2011; Jamshidi Gohari *et al.*, 2013; Vatanpour *et al.*, 2012). Mixed-matric ultrafiltration membranes containing containg inorganic fillers such as silver and zeolite, for instance, can have increased fouling resistance (Hoek *et al.*, 2011). In a recent review of water treatment membrane nanotechnologies Pendergast and Hoek (2011) reported that nanoparticle-containing mixed matrix membranes have the potential to provide enhanced performance, including novel functionalities such as specific adsorption, and improved stability while maintaining the ease of membrane fabrication. Even though such membranes are not yet commercially available, it can be argued that as industrial scale nanoproduction grows, costs will come down. It is hoped that some of these research led improvements will make their way into the open market. While most studies have reported on gas phase purification, it can be reasoned that mixed-matrix membranes due to their specificity and enhanced mechanical strength have great potential for removal of contaminants such as F^-, U and As from wastewater. Zomoza *et al.* (2013) reported the first examples of metal organic framework based mixed-matrix membranes outperforming state-of-art polymers. They noted the high application potential of these composites. Researchers looking at new opportunities should consider studying F^- removal from aqueous solution using mixed-matrix membranes containing various adsorbents/fillers.

6.3 ARSENIC

6.3.1 *Introduction*

Arsenic can be found in two primary forms; organic and inorganic. Organic species of As are mainly found in food, such as shellfish, and include forms as monomethyl arsenic acid (MMAA), dimethyl arsenic acid (DMAA) and arseno-sugars. Inorganic arsenic (i-As) occurs in two valence states, arsenite (As(III)) and arsenate (As(V)). In natural waters, As(III) species consist primarily of arsenious acid (H_3AsO_3) and As(V) species is predominantly present as H_2AsO^{4-} and $HAsO_4^{2-}$

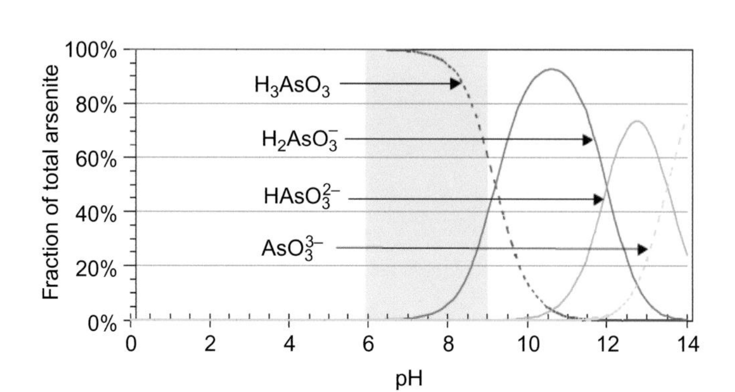

Figure 6.8. Dissociation of arsenite [As(III)] (USEPA, 2003).

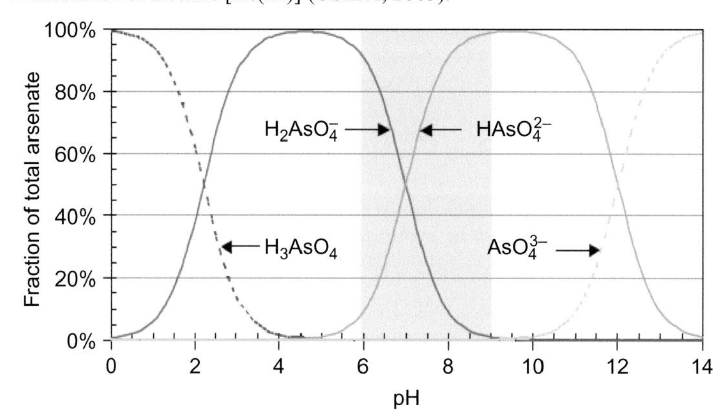

Figure 6.9. Dissociation of arsenate [As(V)] (USEPA, 2003).

(Clifford and Lin, 1995). Most natural waters contain the more toxic inorganic forms of As. Natural groundwater contains predominantly As(III) under reducing conditions whereas As(V) is the principal species under oxidizing conditions.

6.3.2 *Arsenic chemistry*

As is a metal found in the earth's crust, most commonly in the form of iron arsenide sulfide (FeAsS). Arsenic can also be found in the atmosphere as arsenic trioxide dusts, a by-product of industrial smelting operations (http://chemwiki.ucdavis.edu). Through erosion and dissolution, As can enter natural ground and surface waters. Once dissolved, it can take many forms, both organic and inorganic.

Arsenite and arsenate jointly exist in four different forms. The speciation of these molecules changes by dissociation and is pH dependent. The kinetics of dissociation for each are nearly instantaneous. The pH dependencies of arsenite and arsenate are depicted in Figure 6.8 and Figure 6.9, respectively (USEPA, 2003).

Particularly, at any pH less than 9, arsenite will appear as a neutral species H_3AsO_3 where, the neutral form of arsenate H_3AsO_3 is only present at $pH < 3$. This is very important for determining appropriate treatment technologies (Fig. 6.10).

6.3.3 *Technologies for arsenic removal*

As removal technologies discussed in this chapter are grouped into four broad categories: precipitative processes, adsorption processes, ion exchange processes, and pressure-driven membrane processes. At least, one treatment technology was described for each category.

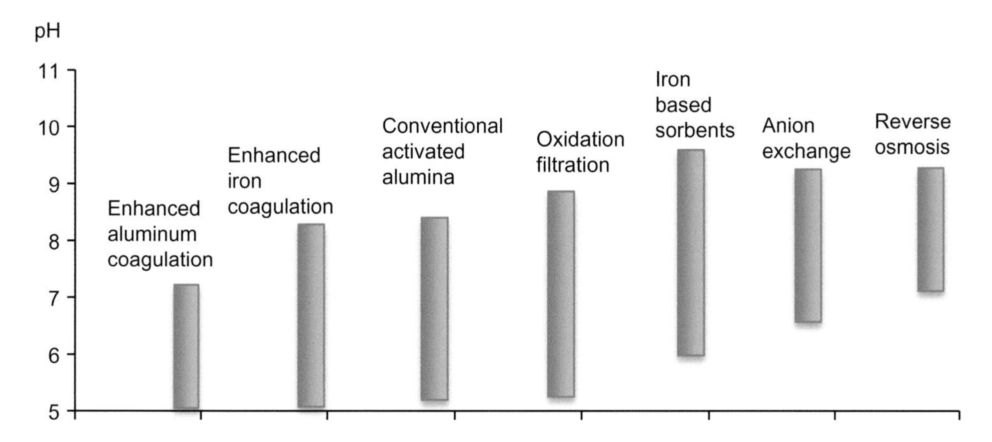

Figure 6.10. Optimal pH ranges for arsenic treatment technologies (adapted from USEPA, 2003).

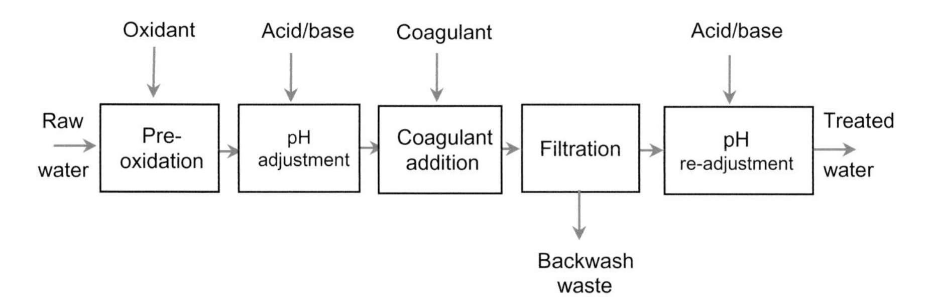

Figure 6.11. Generic coagulation/filtration process flow diagram (adapted from USEPA, 2003).

6.3.3.1 *Coagulation/filtration*

Coagulation filtration is a process based on using an iron or aluminum salt to pretreat water contaminated by As, heavy metals, and dissolved solids. The addition of the salt acts to coagulate the contaminants so that removal using standard filtration can be achieved. This process can be accomplished in large gravity settling basins or in pressure vessels for smaller systems. Major components of a basic coagulation/filtration facility include chemical feed systems, mixing equipment, basins for rapid mix, flocculation, settling, filter media, sludge handling equipment, and filter backwash facilities. Settling may not be necessary in situations where the influent particle concentration is very low. Treatment plants without settling are known as direct filtration plants. As is removed in the pentavalent form, which adsorbs onto coagulated flocs and can be then removed by filtration. As(III) removal during coagulation with alum, ferric chloride, and ferric sulfate has been shown to be less efficient than As(V) under comparable conditions (Edwards, 1994; Gulledge and O'Conner, 1973; Hering *et al.*, 1996; Shen, 1973; Sorg and Logsdon, 1978), thus As(III) has to be previously oxidized (Kartinen and Martin, 1995). Conversion of As(III) to As(V) can be accomplished by providing an oxidizing agent at the head of any proposed As removal process (Fig. 6.11). Chlorine, permanganate and ozone are highly effective for this purpose.

The oxidation-reduction reaction for chlorine, permanganate and ozone are provided in the following equations (USEPA, 2003):

$$H_3AsO_3 + OCl^- \rightarrow H_2AsO_4^- + H^+ + Cl^-$$

$$3H_3AsO_3 + 2MnO_4^- \rightarrow 3H_2AsO_4^- + H^+ + 2MnO_2 + H_2O$$

$$H_3AsO_3 + O_3 \rightarrow H_2AsO_3 \rightarrow H_2AsO_4^- + O_2$$

McNeill and Edwards (1995) reported that Fe and Al based coagulants are mostly used compared to other chemical coagulants. Effective coagulant dosage ranges were 5–25 $mg\,L^{-1}$ of ferric chloride and as much as 40 $mg\,L^{-1}$ of alum (Pallier *et al.*, 2010; USEPA, 2003). Recently, three aluminum based coagulants (aluminum chloride and two types of polyaluminum chloride) were studied by Hu *et al.* (2012). They concluded that each one reduced the concentration of As below the MCL with an initial As(V) concentration of 280 $\mu g\,L^{-1}$. In another study, Ravenscroft *et al.* (2009) found that Fe based coagulants are more effective for water treatment than the Al based coagulants. Unluckily, the production of a large amount of sludge with a considerable concentration of As constitutes the main constraint for the emergence of this treatment technology.

6.3.3.2 *Oxidation and filtration*

Arsenic consists of two major oxyanions, As(III) and As(V) in water (Smedley *et al.*, 2002). These two As species exhibit very different affinities to the mineral surfaces, and the retention of both As(V) and As(III) is strongly pH dependent (Meng *et al.*, 2000) Both inorganic and organic states of the As(III) tend to be more toxic to humans than those of the As(V) forms. The efficiency of most separation methods for As(III) removal is low. To obtain an effective and efficient separation, an oxidation process, which can convert As(III) to As(V), is necessary. Upon efficient oxidation, as a pretreatment, the total removal of As can be effectively improved (Bissen *et al.*, 2003). Different oxidants have been investigated in the As oxidation process, e.g. O_2, O_3, H_2O_2, zero-valent aluminum and iron, activated carbon and manganese (Bissen *et al.*, 2003).

The oxidation efficiency varies when using gaseous oxidants as reported in literature. For example, Frank and Clifford (1986) reported that 8% of As(III) were oxidized within 60 min in solutions purged with pure oxygen. Kim and Nriagu (2000) purged a groundwater containing As(III) with air and pure oxygen and observed that between 54 and 57% of As(III) were oxidized within 5 days. Jiang (2001) noticed that although efficient, utilization of O_3 requires high energy input, thus too expensive for developing countries. Direct application of oxygen sources shows limits in the oxidation efficiency. Consequently, catalysis assisted oxidation, such as MnO_2 coated nanostructured capsules (Criscuoli *et al.*, 2012) was adopted and showed excellent performance. The experimental results show that for the feed with As(III) of 0.1 and 0.3 $mg\,L^{-1}$, complete oxidation was achieved after 3 h and 4 h.

Besides oxidants in the gaseous state, metallic oxidants have also often been employed. Both zero-valent aluminum (ZVAl) (Wu *et al.*, 2007) and iron (Lee *et al.*, 2003) showed preferably oxidation performance for As(III). Leupin *et al.* (2005) reported a positive effect in oxidation of As(III) and a very high removal rate was achieved from an aerated groundwater by filtration through sand and zero-valent iron. A demonstration experiment showed that using 4 oxidation/filtration steps, a 50 $mg\,L^{-1}$ As (total) solution was obtained from a solution of 500 $mg\,L^{-1}$ As(III) solution with an almost complete oxidation.

Iron/manganese oxidation is a commonly used method to treat groundwater Chang *et al.*, 2008; (Driehaus *et al.*, 1995). Hydroxides of metal formed during an oxidation can remove soluble As by a subsequent precipitation or adsorption process. Solar oxidation and removal of As (SORAS) is a simple method that uses irradiation of water with sunlight in PET- or other UV-A transparent bottles to reduce As levels from drinking water from 500 $\mu g\,L^{-1}$ As(III) to 50 $\mu g\,L^{-1}$ As (tot). A typical application of iron oxidation for As(III) was the SONO filter. Inside the filter, there is a top layer composed of coarse river sands (CRS), which is an inactive material as a coarse particulate filter, disperser, flow stabilizer and providing mechanical stability in the filter (Hussam *et al.*, 2007), but generates high concentrations of soluble iron and precipitate as $Fe(OH)_3$ to oxide the groundwater.

Inorganic As(III) species in the feed water is oxidized to As(V) species by the active O_2^-, which is produced by the oxidation of soluble Fe(II) with dissolved oxygen and catalyzed by manganese in the composite iron matrix (CIM). This SONO-filter removes As to concentrations less than

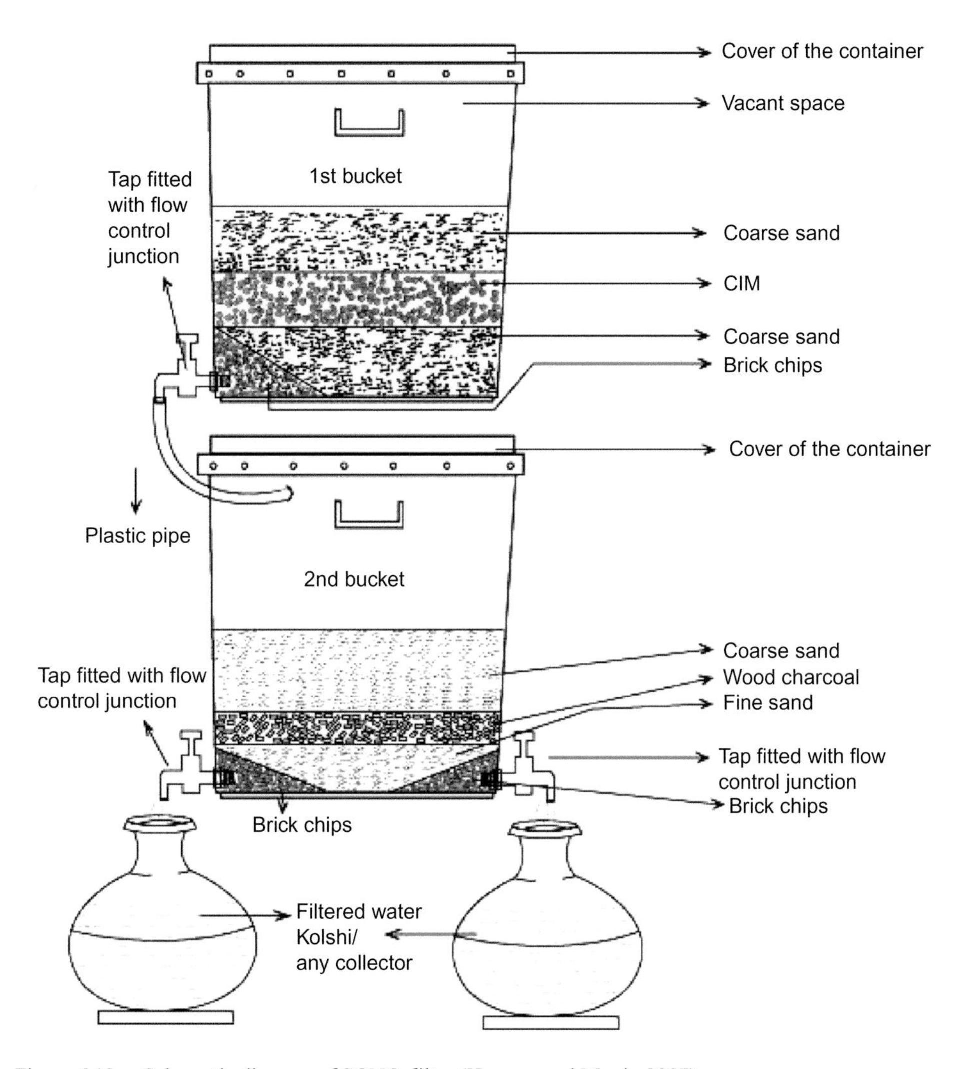

Figure 6.12. Schematic diagram of SONO-filter (Hussam and Munir, 2007).

$14\,\mu g\,L^{-1}$ As (total) until reaching the detection limit ($2\,\mu g\,L^{-1}$) as well as iron, manganese and many other inorganic species to a potable water (Fig. 6.12).

As a summary, As can be effectively removed by a combined oxidation and filtration process. The cost of the oxidants is low and easily accessible. However, as a simple coarse depth filtration process, the removal rate of the filtration step strongly depends on the saturation of the filter, and the concentration of the feed streams. An absolute separation approach is essential to achieve a stable and much better separation performance. Nonetheless, it is clear that using iron and manganese chemicals is practical for the oxidation of As(III) to As(V). Better performance may be achieved if a better separation technology is available.

6.3.3.3 *Adsorptive processes*

Conventional adsorbents such as alumina, iron oxide, manganesia, titania, and ferric phosphate have been studied extensively to remove As from water (Mohan *et al.*, 2007). Absorption is a physical-chemical process by which the adsorbates (ions of targeted solutes) are adsorbed to the surface of an adsorbent. Cupric oxide and iron oxide adsorbers for instance have been investigated

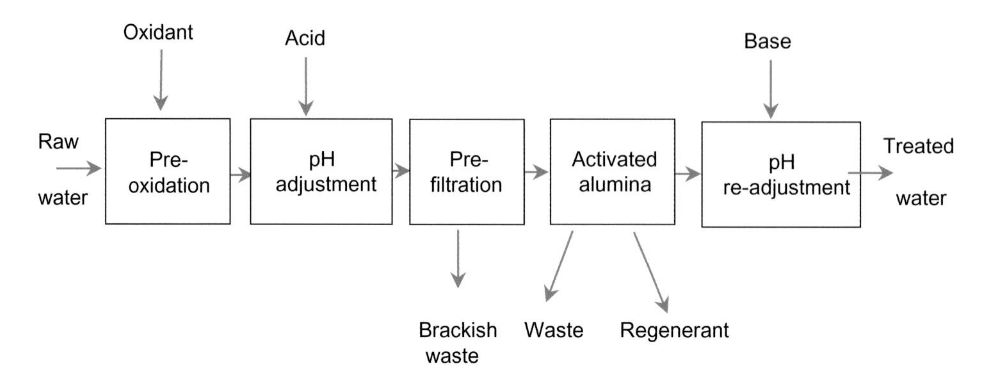

Figure 6.13. Activated alumina process flow diagram (adapted from USEPA, 2003).

as low cost alternatives for rural areas and mobile applications (Chen *et al.*, 2013; Reddy *et al.*, 2013). Rapid small-scale column tests for Arsenate Removal (SCT) in iron oxide packed bed columns have been assessed by Westerhoff *et al.* (2005). It was reported that a simulated 2.5 min empty-bed contact time (EBCT), a model water (pH = 8.6) had q column values of 0.99 to 1.5 mgAs/gGFH *vs.* 0.02 to 0.28 mgAs/gGFH with a comparable pH and EBCT in a natural groundwater indicating a high adsorption efficiency of the SCT.

Fixed-bed filters have been successfully applied for the removal of As in the developing world (Bissen *et al.*, 2003; Sperlich *et al.*, 2005). These filters are simple to operated, feasible for small scale requirements, cost-effective, and normally have low maintenance. Additionally, no dosing of chemicals is required and the amount of residuals is low when adsorbents with high adsorption capacities are used.

Point-of-use (POU) filters have been developed as well (Gurian *et al.*, 2002). The replacement or regeneration frequency of the adsorbents can be minimized if the filtrate is only used for drinking water purposes (Petrusevski *et al.*, 2002). However, the synthesis of most adsorbents, especially efficient adsorbents, are complex, and their performance of re-use is poor, and this process will produce large amounts of As sludge or As containing solid wastes, and other treatment methods are still much desired.

In this section we present two types of commonly used adsorbents, i.e., activated alumina and activated carbon. Activated alumina (AA) is a granulated form of aluminum oxide. It is commonly used to remove As (Guan *et al.*, 2009; Lin *et al.*, 2001). Typically, aluminum oxide (Al_2O_3) granules with a very high internal surface area, in the range of 200 to 300 $m^2\,g^{-1}$ are utilized after a first oxidation step. The As removal efficiency of typically >95% is achieved with a raw water containing arsenite. AA adsorption is a physical/chemical process in which ions in solution are removed on the oxide surface. Feed water is passed continuously through one or more activated alumina beds. Periodically, the activated alumina medium is backwashed to remove any solids that have accumulated in the system. When all available sites are occupied, the activated alumina medium must either be regenerated with a strong base or disposed of entirely. Figure 6.13 shows a typical process flow diagram for Activated Alumina (USEPA, 2003).

The pH has a significant effect on As removal with AA. A pH of 8.2 is significant because it is the "zero point charge" for AA. Below this pH, AA has a net positive charge resulting in a preference for adsorption of anions, including As (USEPA, 2000). Acidic pH levels are generally considered optimum for As removal with AA. The level of competing ions affects the performance of AA for As(V) removal. The following selectivity sequence has been established for AA adsorption (USEPA, 2000):

$$OH^- > H_2AsO_4^- > Si(OH)_3O^- > F^- > HSeO_3^- > TOC > SO_4^{2-} > H_3AsO_3$$

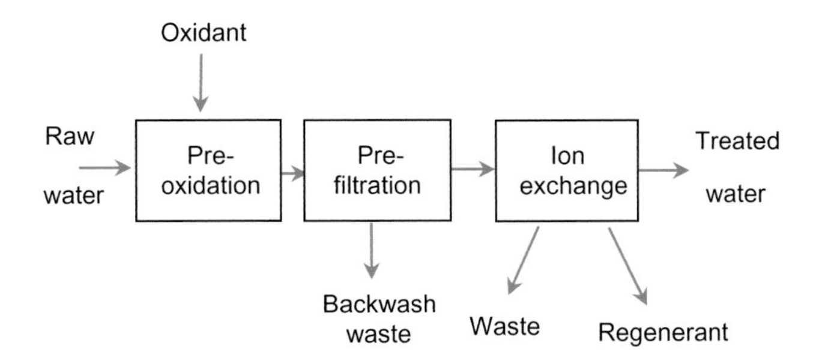

Figure 6.14. Ion exchange process flow diagram (adapted from USEPA, 2003).

The selectivity of AA towards As(III) is poor; therefore, pre-oxidation of As(III) to As(V) is critical.

Activated carbon (AC) as powdered or granulated forms can be used for effective removal of As(III) and As(V) ions from aqueous solutions. AC is a good candidate for the support of the hybrid adsorbents due to the low cost and wide range of available properties (Fierro *et al.*, 2009; Lorenzen *et al.*, 1995; Vitela-Rodriguez *et al.*, 2013). The amount of As uptake by AC is greatly dependent on pH and its oxidation state.

Adsorption of As by AC can be improved by impregnation of AC employing suitable chemicals (e.g., sulfur contain textile dyes for As(III) and Fe(III) salts for removal of As(V) from aqueous solutions (Ansari, 2007). Recently, an interesting study was carried out by Alma *et al.* (2013) on the As removal from water using activated carbon (AC). The authors tested different activated carbons modified with iron hydroxide for their ability to adsorb As from water. They concluded that iron modified activated carbons are efficient adsorbents for As at concentrations lower than 300 $\mu g\, L^{-1}$. According to Mohan and Pittman (2007) Activated carbon can remove 60% As(V) and As(III) but this removal percentage of As is not sufficient to reach drinking water quality. However, sedimentation or filtration processes are required in case of powdered activated carbons, which in turn adds extra cost to the technology.

6.3.3.4 *Ion exchange*

Ion exchange is also frequently used as a treatment technology for arsenic removal. As contaminated water is passed through the resin, contaminant ions are exchanged for other ions in the resin. Ion exchange is often preceded by treatments such as filtration to remove organics, suspended solids, and other contaminants that can foul the resins and reduce their effectiveness. Ion-exchange resins must be periodically regenerated to remove the adsorbed contaminants and replenish the exchanged ions. Regeneration water and spent resin containing high levels of As would require additional treatment prior to disposal or reuse. Alternatively, single-use, non-regenerable ion exchange resins may be used. Figure 6.14 shows a typical process flow diagram for ion exchange.

Before passing As contaminated water, the resin bed are usually flushed with HCl so as to implant labile Cl^- on the surface of the resin, which is later easily exchanged with As. Thus, the effluent contains a considerable amount of Cl^- and additional secondary treatment is needed to improve the quality (Mondal *et al.*, 2013). Since arsenite usually exists as a neutral molecule and is not exchanged, oxidation of As(III) to As(V) is an important pretreatment step for ion exchange processes (Kartinen and Martin, 1995).

The exchange affinity of various ions is a function of the net surface charge. Therefore, the efficiency of the IX process for As(V) removal depends strongly on the solution pH and the concentration of other anions, most notably sulfates and nitrates. These and other anions compete for sites on the exchange resin according to the following selectivity sequence (Clifford, 1999).

Korngold *et al.* (2001) used strong basic anion-exchange resins for the removal of As(V). The resin was regenerated with NaCl or HCl. More than 99% of As was removed by the resins at an initial As concentration of 600 $\mu g L^{-1}$. An anion exchanger (AE) prepared from coconut coir pith (CP) was applied for the removal of As(V) from aqueous solutions (Anirudhan *et al.*, 2007) and a maximum removal of 99.2% was obtained for an initial concentration of 1 $mg L^{-1}$ As(V) at pH 7.0 at an adsorbent dose of 2 $g L^{-1}$. Regeneration of the IE is an issue, particularly that a large amount of water is required to rinse the system.

Phytoremediation has received increasing attention after the discovery of hyperaccumulating plants which are able to accumulate, translocate and concentrate high amounts of certain toxic elements in their above-ground/harvestable parts (Rahman *et al.*, 2011). Phytoremediation of contaminated water by aquatic macrophytes would be a good option in the long term. A large number of aquatic plant species have been tested for the remediation of toxic elements from fresh water systems. Few aquatic plants (mostly macrophytes) have shown the ability to accumulate high levels of As from water. If the term to achieve the desired effect is short enough, the phytoremediation will be an effective As removal process with low energy consumption. The biological process is normally a slow and less efficient process. It is most probably more practical to combine both phytoremediation and other high efficient separation process to remove As.

6.4 URANIUM

6.4.1 *Introduction*

Uranium because of its radioactivity and heavy-metal toxicity is highly lethal not only with respect to human health but also to the whole ecological system (Zou *et al.*, 2009). While dissolved U usually occurs in most natural waters at very low concentrations, U mining, milling, processing, enriching, and disposal all contribute to contaminate surface water and groundwater. Significant amounts of U have also found their way into the environment through the actions associated with the nuclear industry (Ghasemi *et al.*, 2011). Hence, the removal of U from wastewater is important not only for the nuclear industry, but also for environmental remediation. Investigation on separation of U from wastewater is thus vital.

Pollutants from the ore-processing industries, surface and underground mines may find their way into the groundwater in a variety of ways. It can be seen from this scheme, that pollution of surface and ground waters, as well as air takes place during mining. Adsorption by low cost adsorbents, for example, provides an environmentally and economically friendly technique for removing U from wastewaters (Iakovleva and Sillanpää, 2013).

6.4.2 *Biosorption of uranium by algae biomass*

The search for economical and eco-friendly solutions for U removal from water has led to the utilization of biological materials (e.g., microbial and plant origin) as adsorbents since they interact effectively with heavy metals (Ghasemi *et al.*, 2011). Biosorption is described by the removal of heavy metals by dead biomass from aqueous solutions and is attributed mainly to the ligands present in their cell wall biomolecules. There are many natural adsorbents such as marine algae, bacteria, fungi, and industrial wastes that have been used for U removal from water solutions (Bayramoglu *et al.*, 2006; Kalin *et al.*, 2005; Khani *et al.*, 2008; Li *et al.*, 2004; Parab *et al.*, 2005). Among these, algal biomass has received abundant attention due to its low cost, environmental friendliness, and elevated adsorption capacity.

Uranium (VI) biosorption by Ca-pretreated *Cystoseira indica* biomass was studied by Ghasemi *et al.* (2011) using a continuous packed bed column. Metal uptake capacity was found to remain constant with the rise in bed height, while the breakthrough and the exhaustion times increased. Moreover, a decrease in the column bed height resulted in a lower percentage metal removal. It was found that the adsorption breakthrough was strongly dependent on the liquid flow rate, as expected. The authors argued that a successful biosorption process operation requires the multiple reuses of the sorbent, which would greatly reduce the process cost as well as decreasing the dependency of the process on continuous supply of the sorbent. The results of Ghasemi *et al.*

Figure 6.15. SEM micrograph of sample: (a) zeolites; (b) manganese oxide coated zeolites (Zou *et al.*, 2006).

(2011) on column regeneration for three cycles indicated that the usability of *C. indica* biomass for hexavalent uranium (U(VI)) removal and recovery is viable. They concluded that high biosorption efficiency of the alga, low biomass cost, less dependency on the biomass due to reuse, and high efficient elutant make this process an effective, cheap, and alternative technique for treatment of U(VI) bearing solutions.

In a related study, Khani (2011) reported on the removal of U ions from aqueous solutions using *Padina sp.*, a brown marine algal biomass. Four main parameters (pH and initial U concentration in solutions, contact time and temperature) were assessed on U uptake. Results showed that the adsorption data adequately fitted a second-order polynomial model. The optimum pH and initial U concentration in solutions, contact time and temperature were found to be 4,778 mg L^{-1}, 74 min, and 37°C, respectively. The maximum U uptake was predicted and experimentally validated. The maximum monolayer adsorption capacity was found to be as high as 376 mg g^{-1}.

6.4.3 *Removal of uranium (VI) from water using zeolite coated with manganese oxide*

Natural zeolites have great potential for heavy metal removal due to their ion exchange ability (Zou *et al.*, 2009). The physical structure of zeolites is highly porous, with interconnected cavities, in which the metal ions and water molecules can interact with the zeolite surface (Fig. 6.15). SEM photographs in Figure 6.15 (a and b) were taken by Zou *et al.* (2006) at 10000×, 5000× magnifications to observe the surface morphology of zeolites and manganese oxide coated zeolites, respectively. At the micron scale, the synthetic coating is composed of small particles on top of a more consolidated coating. SEM images indicated a much rougher surface after the manganese oxide coating. In studies with Cu(II) and Pb(II) Zou *et al.* (2006) showed that adsorption is a spontaneous and endothermic process with a rise in temperature favoring the adsorption.

Manganese oxides, with high adsorptive property, are usually considered as the most significant foragers for trace metals in soil, sediments, and rocks (Zou *et al.*, 2009). The surface charge of manganese oxides is usually negative, so that they can be used as adsorbents to remove heavy metal ions from wastewater. However, pure manganese oxide as a filter medium is not favorable for both economic reasons and unfavorable physical and chemical characteristics, but the coating of manganese oxide on a medium surface such as provided by zeolites may be promising for heavy metal removal from wastewater (Han *et al.*, 2007).

In a study by Han *et al.* (2007), manganese oxide coated zeolites (MOCZ) were synthesized and the adsorption properties for U(VI) by MOCZ were investigated. Their study clearly established that MOCZ is an effective adsorbent for U(VI) removal from aqueous solutions. They reported that the U(VI) binding capacity by MOCZ was strongly dependent on the initial pH, initial U(VI) concentration and temperature. It was noted that an increase in the temperature resulted in a higher metal loading per unit weight of MOCZ. Adsorption capacity increased slightly with increasing temperature. The equilibrium sorption of U(VI) was determined from the Langmuir equation and found to be 15 mg g^{-1} at 293 K and pH 4.0. The authors concluded that the thermodynamics of the U(VI) ion/MOCZ system indicates spontaneous and endothermic nature of the process. This is a similar conclusion as that reached by Zou *et al.* (2006) with zeolite coated with manganese oxide; adsorption is a spontaneous and endothermic process with a rise in temperature favoring the adsorption.

6.4.4 *Removal of uranium from groundwater using biochar, carbonaceous adsorbents and magnetic composite particles*

Kumar *et al.* (2011) argued that the ever-increasing growth of bio-refineries is expected to produce huge amounts of lignocellulose biochar as a by-product. Hydrothermally produced biochar is a porous and amorphous solid rich in active functional groups (e.g., hydroxyl/phenolic, carboxylic, and carbonyl groups). This by-product has great potential for use as an inexpensive adsorbent for heavy metal removal from wastewater. Uranium (VI) removal from groundwater, for example, was assessed by Kumar *et al.* (2011) using biochar produced from hydrothermal carbonization. A batch adsorption experiment at the natural pH (about 3.9) of biochar indicated an H-type isotherm with an adsorption capacity estimated at 2 mg of U g^{-1} of biochar. The adsorption process was highly dependent on the pH of the system. An increase towards neutral pH resulted in a maximum adsorption of 4 mg U g^{-1} of biochar. The authors concluded that the adsorption of U onto biochar is an attractive alternative to treat U(VI)-contaminated groundwater. Compared to other remediation strategies, the feasibility of biochar as U(VI) adsorbent is supported by its environmentally benign nature. The major advantage of biochar is that it could serve as an effective and green adsorbent for U without causing environmental damage.

A low-cost and highly-efficient adsorbent (HTC-COOH) functionalized with carboxylic groups was produced by Liu *et al.* (2013) through oxidizing hydrothermal carbon by HNO_3 solution. Adsorption studies showed that the amounts of U(VI) adsorbed by hydrothermal carbon (HTC) and HTC-COOH were strongly pH-dependent, and increased with the initial concentration of U(VI) and temperature. The adsorption of U(VI) onto HTC and HTC-COOH was well-described by Langmuir isothermal equations and pseudo-second order kinetics models. Calculated monolayer adsorption capacity increased from 62 to 205 mg g^{-1} after carboxylation. There was also increased selectivity. Liu *et al.* (2013) investigated the adsorption of U(U(VI)) from aqueous solution, in the presence of other ions (i.e., Mg^{2+}, Co^{2+}, Ni^{2+}, Zn^{2+} and Mn^{2+}). The results shown in Figure 6.16 indicated that while HTC showed no selectivity to U(VI), the amount of U(VI) adsorbed on HTC-COOH (15 mg g^{-1}) was much higher than for the other ions. The authors concluded that HTC-COOH is an inexpensive and excellent adsorbent for selective removal of U(VI) in aqueous solution.

In a related study with similar selectivity results, Fan *et al.* (2012) reported on the use of magnetic Fe_3O_4@SiO_2 composite particles as an effective adsorbent material for the removal of U(VI) from aqueous solution. The results shown in Figure 6.17, suggest that the sorption efficiency of U was still high (about 93%) in the presence of other cations. The authors concluded that the presence of other ions had almost no effect on the sorption of U(VI) on the magnetic composite particles. This is a similar result to that obtained by Liu *et al.* (2013) working with a high-efficient adsorbent (HTCCOOH) functionalized with carboxylic groups (Fig. 6.16); the amount of U(VI) adsorbed on HTC-COOH was much higher than for the other ions. Fan *et al.* (2012) concluded that magnetic Fe_3O_4@SiO_2 composite particles could be a perfect candidate as an adsorbent to remove the toxic and radioactive U(VI) from solution.

6.4.5 *Uranium removal from water using ultrafiltration, reverse osmosis, nanofiltration and electrodialysis*

A pilot plant was built by Montaña *et al.* (2013) to assess the effectiveness of ultrafiltration (UF), reverse osmosis (RO), and electrodialysis reversal (EDR) in improving the quality of the

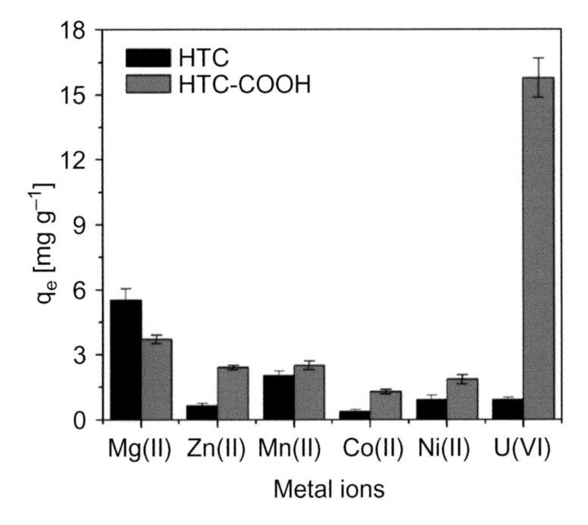

Figure 6.16. Competitive sorption capacity of cations (Mg^{2+}, Co^{2+}, Ni^{2+}, Zn^{2+}, Mn^{2+} and U^{6+}), pH = 4.0, t = 60 min, $C_0 = 5\,mg\,L^{-1}$, $V = 150\,mL$, $m = 0.02\,g$ and $T = 308.15\,K$ (Liu *et al.*, 2013).

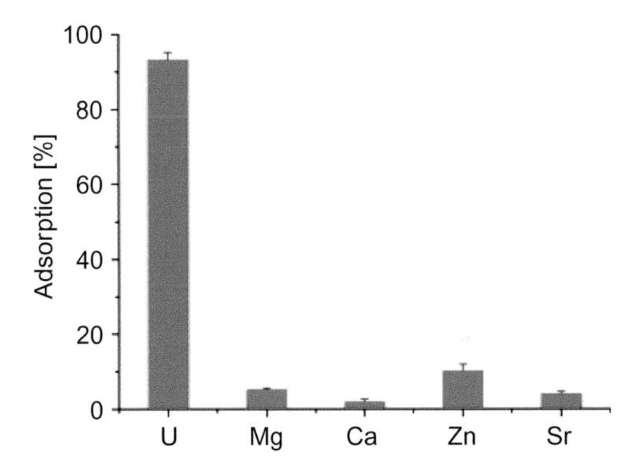

Figure 6.17. Competitive sorption of cations onto Fe_3O_4@SiO_2 magnetic composites (50 mg L^{-1} for every ions, pH = 6.0, Time = 180 min, $m/v = 2.5\,g\,L^{-1}$ (Fan *et al.*, 2012).

water supplied to the Barcelona metropolitan area from the Llobregat River in Spain. Their paper presented results from two studies to reduce natural radioactivity. The results from the pilot plant with four different scenarios were used to design a full-scale treatment plant (Fig. 6.18 and Fig. 6.19). Samples taken at different steps of the treatment were analyzed to determine gross alpha, gross beta and U activity. The results obtained revealed a significant improvement in the radiological water quality provided by both membrane techniques (i.e., RO and EDR showed removal rates higher than 60%). However, UF did not show any significant removal capacity for gross alpha, gross beta or U activities. RO was better at reducing the radiological parameters studied and this treatment was selected and applied for the full scale treatment plant. The RO treatment used at the full-scale plant reduced the concentration of both gross alpha and gross beta activities and also produced water of high quality with an average removal of 95% for gross alpha activity and almost 93% for gross beta activity at the treatment plant (Fig. 6.19).

In a related study, Khedr (2013) assessed the removal of radium, U, as uranyl cation, or carbonate complexes, and radon by reverse osmosis (RO) and nanofiltration (NF) in comparison with the conventional methods of ion exchange resins (IERs), chemical precipitation/softening, coagulation, and adsorption on surface active media. IERs and chemical softening achieved radionuclide rejection from 32 to 95%, but with loss of process efficiency due to undesired

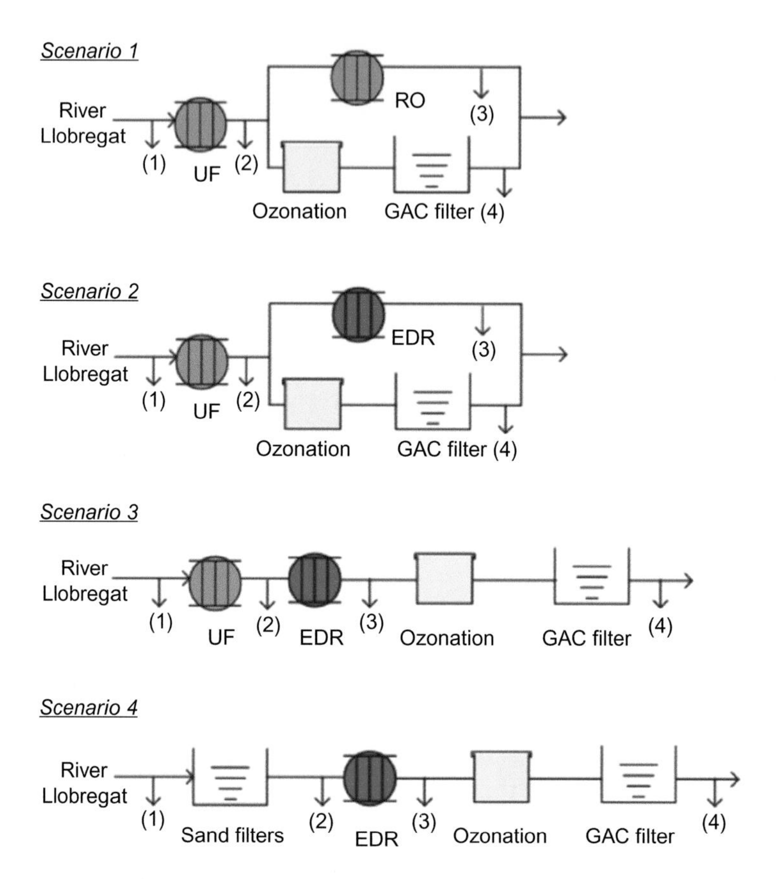

Figure 6.18. Treatment layouts for the four scenarios studied in a pilot plant for removing radionuclides from drinking water. Sampling points are represented by numbers in brackets (Montaña *et al.*, 2013). Code: ultrafiltration (UF), reverse osmosis (RO), electrodialysis reversal (EDR).

parallel removal of similar ions. Removal by IERs was too dependent on the resin form and water pH and required periodical shutdown for regeneration of resin which was slow and seldom complete. Softening required chemical dosing stoichiometric to isotope removal, disposal of contaminated sludge and subsequent water filtration. Coagulation failed to remove radium. In the other hand, the removal of U ranged from zero to 93% depending on pH due to formation of different U complexes. Only RO, parallel to water desalination, showed steady, high rejection of all isotopes (i.e., up to 99%) without interference of similar ions, regeneration, or subsequent filtration. NF showed similar behavior, but with lower water desalination efficiency.

As a final example, Villalobos-Rodriguez *et al.* (2012) reported on the use of ultrafiltration for removal of U from water, with composite activated carbon cellulose triacetate membranes. Uranium removal was found to be $35 \pm 7\%$. Results suggested that co-adsorption of U and iron by the carbon cellulose triacetate membranes during filtration, as the leading rejection path. It can be argued that the U removal reported by Villalobos-Rodriguez *et al.* (2012) was not as good as that presented by Montaña *et al.* (2013) and Khedr (2013) for reverse osmosis membranes (i.e., greater than 90% removal of U).

6.4.6 *Potential of using mixed matrix membranes for removal of uranium from water*

Zomoza *et al.* (2013) reported the first examples of metal organic framework based mixed-matrix membranes outperforming state-of-art polymers. They noted the high application potential of

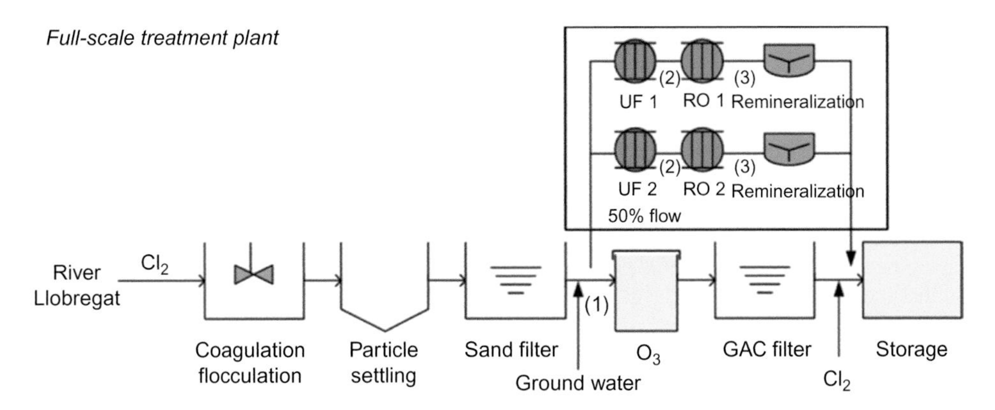

Figure 6.19. Schematic representation of the full-scale treatment plant at Llobregat River in Barcelona Spain. Sample points represented by numbers are in brackets (Montaña *et al.*, 2013). Code: ultrafiltration (UF), reverse osmosis (RO).

these composites. Researchers looking at new opportunities should consider studying U removal from aqueous solution using mixed-matrix membranes containing various adsorbents/fillers such as zeolites. Natural zeolites have great applicability for heavy metal elimination due to their ion exchange ability (Zhou *et al.*, 2009).

Mixed matrix membranes, a somewhat new development, take advantage of both the relatively low cost of fabrication of polymeric membranes and the mechanical strength and functional properties of inorganic materials, such as zeolite, silver, silica as well as nanoparticles (Hoek *et al.*, 2011; Jamshidi Gohari *et al.*, 2013; Vatanpour *et al.*, 2012). Incorporation of the nanoparticles into a nanofiltration coating layer may combine the oxidation and separation of heavy metals in one step as illustrated in Figure 6.14. In a recent review of water treatment membrane nanotechnologies Pendergast and Hoek (2011) reported that nanoparticle-containing mixed matrix membranes have the potential to provide enhanced performance, including novel functionalities such as specific adsorption, and improved stability while maintaining the ease of membrane fabrication. It can be argued that mixed-matrix membranes due to their specificity and enhanced mechanical strength have great potential for removal of contaminants such as F^-, U and As from wastewater.

6.5 CONCLUDING REMARKS AND OUTLOOK

Various treatment methods based on conventional, modern and hybrid technologies have been applied for remediation of F^-, U and As in many parts of the world. These techniques have been critically reviewed in this chapter. Metal organic framework based mixed-matrix membranes have been reported to outperform state-of-art polymers. These composite membranes containing various adsorbents/fillers such as zeolites have high application potential and should be studied further for removal of heavy metals from wastewater.

There is a need for low-cost and proven technologies that can effectively treat polluted water especially in developing countries. Membrane technology has already been successfully used on large-scale for removal of inorganic anions such as nitrate, as well as F^-, U and As. Two disadvantages though are membrane fouling and concentrated brine discharge management and/or treatment. While novel technologies such as membrane bioreactors allow for complete contaminant removal, process operations are insufficiently stable, and also limited due to economic reasons. The major challenges therefore are the design of more efficient membranes. Cost-effective operating conditions are required, especially for long-term processes without or with minimal membrane inorganic and/or biological fouling, and a reduction in the energy consumption requirements.

The applicability of chitin, chitosan and chemically modified chitosan as adsorbents should also be investigated further for the removal of F^-, As and U from drinking water. We can expect this to be one area of rapid development. Chitosan which is derived from chitin one of the mainly components of crustacean shells of prawn, crab, shrimp or lobster, is inexpensive and versatile and has the ability to coordinate various metal ions.

In closing, F^-, As and U removal from wastewater using adsorbent materials and integrated membrane systems are rapidly evolving areas. Mixed matrix membranes containing adsorbent fillers/nanoparticles may be the next stage of growth. It should be kept in mind though that any new process or technology that is produced must be relatively inexpensive and simple to operate so that it can be employed in the developing world where the need is often most acute. This will be the main challenge facing us.

REFERENCES

Alma, V., Vitela, R. & Jose, R. (2013) Arsenic removal by modified activated carbons with iron hydro(oxide) nanoparticles. *Journal of Environmental Management*, 114, 225–231.

Al-Obeidani, S., Al-Hinai, H., Goosen, M.F.A., Sablani, S., Taniguchi, Y. & Okamura, H. (2008) Chemical cleaning of oil contaminated polyethylene hollow fiber microfiltration membranes. *Journal of Membrane Science*, 307, 299–308.

Anirudhan, T.S. & Unnithan, M.R. (2007) Arsenic (V) removal from aqueous solutions using an anion exchanger derived from coconut coir pith and its recovery. *Chemosphere*, 66, 60–66.

Ansari, R. & Sadegh, M. (2007) Application of activated carbon for removal of arsenic ions from aqueous solutions. *E-Journal of Chemistry*, 4 (1), 103–108.

Ayoob, S., Gupta, A.K. & Bhat, V.T. (2008) A conceptual overview on sustainable technologies for defluoridation of drinking water and removal mechanisms. *Critical Reviews in Environmental Science and Technology*, 38 (6), 401–470.

Bayramoglu, G., Celik, G. & Arica, M.Y. (2006) Studies on accumulation of uranium by fungus *Lentinus sajor-caju*. *Journal of Hazardous Materials*, 136, 345–353.

Bhatnagar, A., Kumar, E. & Sillanpää, M. (2011) Fluoride removal from water by adsorption: a review. *Chemical Engineering Journal*, 171, 811–840.

Bissen, M. & Frimmel, F. (2003) Arsenic – a review. Part I: Occurrence, toxicity, speciation, mobility. *Acta Hydrochimica et Hydrobiologica*, 31 (1), 9–18.

Chang, J.S., Kim, Y.H. & Kim, K.W. (2008) The ars genotype characterization of arsenic-resistant bacteria from arsenic-contaminated gold-silver mines in the Republic of Korea. *Applied Microbiology and Biotechnology*, 80, 155–165.

Chen, N., Zhang, Z., Feng, C., Zhu, D., Yang, Y. & Sugiura, N. (2011) Preparation and characterization of porous granular ceramic containing dispersed aluminum and iron oxides as adsorbents for fluoride removal from aqueous solution. *Journal of Hazardous Materials*, 186, 863–868.

Chen, Y., Wu, F., Liu, M., Parvez, F., Slavkovich, V., Eunus, M., Ahmed, A., Argos, M., Islam, T., Rakibuz-Zaman, M., Hasan, R., Sarwar, G., Levy, D., Graziano, J. & Ahsan, H. (2013) A prospective study of arsenic exposure, arsenic methylation capacity, and risk of cardiovascular disease in Bangladesh. *Environmental Health Perspectives*, 121, 832–838.

Clifford, D. & Lin, C.C. (1995) Ion exchange, activated alumina, and membrane processes for arsenic removal from groundwater. *Proceedings of the 45th Annual Environmental Engineering Conference, February 1995, University of Kansas, KS, USA*.

Clifford, D.A. (1999) Ion exchange and inorganic adsorption. *Water Quality Treatment*, 4, 561–564.

Criscuoli, A., Majumdar, S., Figoli, A., Sahoo, G.C., Bafaro, P., Bandyopadhyay, S. & Drioli, E. (2012) As (III) oxidation by MnO_2 coated PEEK-WC nanostructured capsules. *Journal of Hazardous Materials*, 211, 281–287.

Dolar, D., Košutić, K. & Vučić, B. (2011) RO/NF treatment of wastewater from fertilizer factory – removal of fluoride and phosphate. *Desalination*, 15, 237–241.

Driehaus, W., Seith, R. & Jekel, M. (1995) Oxidation of arsenate (III) with manganese oxides in water treatment. *Water Research*, 29, 297–305.

Edwards, M.A. (1994) Chemistry of arsenic removal during coagulation and Fe-Mn oxidation. *Journal American Water Works Association*, 64–77.

Fan, F.Y., Qin, Z., Bai, J., Rong, W.D., Fan, F.Y., Tian, W., Wu, X.L., Wang, Y. & Zhao, L. (2012) Rapid removal of uranium from aqueous solutions using magnetic Fe_3O_4@SiO_2 composite particles. *Journal of Environmental Radioactivity*, 106, 40–46.

Fierro, V., Muñiz, G., Gonzalez-Sánchez, G., Ballinas, M.L. & Celzard, A. (2009) Arsenic removal by iron-doped activated carbons prepared by ferric chloride forced hydrolysis. *Journal of Hazardous Materials*, 168, 430–437.

Frank, P. & Clifford, D.A. (1986) Arsenic (3) oxidation and removal from drinking water. Volume 86 (158607), Water Engineering Research Laboratory, Office of Research and Development, US Environmental Protection Agency.

Ghasemi, M., Keshtkar, A.R., Dabbagh, R. & Safdari, S.J. (2011) Biosorption of uranium(VI) from aqueous solutions by Ca-pretreated *Cystoseira indica* alga: breakthrough curves studies and modelling. *Journal of Hazardous Materials*, 189, 141–149.

Ghorai, S. & Pantk, K. (2005) Equilibrium, kinetics and breakthrough studies for adsorption of fluoride on activated alumina. *Separation and Purification Technology*, 42, 265–271.

Goosen, M.F.A., Sablani, S., Del-Cin, M. & Wilf, M. (2011) Effect of cyclic changes in temperature and pressure on permeation properties of composite polyamide spiral wound reverse osmosis membranes. *Separation Science and Technology*, 46 (1), 14–26.

Guan, X., Dong, H., Ma, J. & Jiang, L. (2009) Removal of arsenic from water: effects of competing anions on As (III) removal in $KMnO_4$Fe (II) process. *Water Research*, 43, 3891–3899.

Gulledge, J.H. & O'Connor, J.T. (1973) Removal of arsenic (V) from water by adsorption on aluminum and ferric hydroxides.*Journal American Water Works Association*, 8, 548–552.

Gurian, P.L. & Small, M.J. (2002) Point-of-use treatment and the revised arsenic MCL. *Journal American Water Works Association*, 94, 101–108.

Han, R., Zou, W., Wang, Y. & Zhu, L. (2007) Removal of uranium(VI) from aqueous solutions by manganese oxide coated zeolite: discussion of adsorption isotherms and pH effect. *Journal of Environmental Radioactivity*, 93, 127–143.

Hering, J.G. & Elimelech, M. (1996) Arsenic removal by enhanced coagulation and membrane processes. American Water Works Association, Denver, CO.

Hoek, E.V.M., Ghosh, A.K., Huang, X., Liong, M. & Zink, J.I. (2011) Physical-chemical properties, separation performance, and fouling resistance of mixed-matrix ultrafiltration membranes. *Desalination*, 283, 89–99.

Hu, C.Y., Lo, S.L. & Kuan, W.H. (2003) Effects of co-existing anions on fluoride removal in electrocoagulation (EC) process using aluminum electrodes. *Water Research*, 37, 4513–4523.

Hussam, A. & Munir, A.K.M. (2007) A simple and effective arsenic filter based on composite iron matrix: development and deployment studies for groundwater of Bangladesh. *Journal of Environmental Science and Health*, Part A: *Toxic/Hazardous Substances and Environmental Engineering*, 42, 1869–1878.

Iakovleva, E. & Sillanpää, M. (2013) The use of low-cost adsorbents for wastewater purification in mining industries. *Environmental Science and Pollution Research*, 20 (11), 7878–7899.

Jamshidi Gohari, R., Lau, W.J., Matsuura, T., Halakoo, E. & Ismail, A.F. (2013) Adsorptive removal of Pb(II) from aqueous solution by novel PES/HMO ultrafiltration mixed matrix membrane. *Separation and Purification Technology*, 120, 59–68.

Jiang, X. (2001) Arsenic trioxide induces apoptosis in human gastric cancer cells through up-regulation of P53 and activation of caspase-3. *International Journal of Cancer*, 91 (2), 173–179.

Kalin, M., Wheeler, W.N. & Meinrath, G. (2005) The removal of uranium from mining waste water using algal/microbial biomass. *Journal of Environmental Radioactivity*, 78, 151–177.

Kamble, S.P., Jagtap, S., Labhsetwar, N.K., Thakare, D., Godfrey, S., Devotta, S. & Rayalu, S.S. (2007) Defluoridation of drinking water using chitin, chitosan and lanthanum-modified chitosan. *Chemical Engineering Journal*, 129, 173–180.

Kartinen Jr., E.O. & Martin, C.J. (1995) An overview of arsenic removal processes. *Desalination*, 103, 79–88.

Khani, M.H. (2011) Statistical analysis and isotherm study of uranium biosorption by *Padina* sp. algae biomass. *Environmental Science and Pollution Research*, 18 (5), 790–799.

Khani, M.H., Keshtkar, A.R., Ghannadi, M. & Pahlavanzadeh, H. (2008) Equilibrium, kinetic and thermodynamic study of the biosorption of uranium onto *Cystoseria indica* algae. *Journal of Hazardous Materials*, 150, 612–618.

Khatibikamala, V., Torabiana, A., Janpoora, F. & Hoshyaripourb, G. (2010) Fluoride removal from industrial wastewater using electrocoagulation and its adsorption kinetics. *Journal of Hazardous Materials*, 179, 276–280.

Khedr, M.G. (2013) Radioactive contamination of groundwater, special aspects and advantages of removal by reverse osmosis and nanofiltration. *Desalination*, 321, 47–54.

Kim, M. & Nariagu, J. (2000) Oxidation of arsenite in groundwater using ozone and oxygen. *Science of the Total Environment*, 247, 71–79.

Korngold, E., Belayev, N. & Aronov, L. (2001) Removal of arsenic from drinking water by anion exchangers. *Desalination*, 141, 81–84.

Kowalchuck, E.E. (2011) *Selective fluoride removal by aluminum precipitation & membrane filtration*. Master's Thesis. University of New Mexico, Albuquerque, NM. Available from: http://repository.unm.edu/handle/1928/17436 [accessed March 2014].

Kumar, S., Loganathan, V.A., Gupta, R.B. & Barnett, M.O. (2011) An assessment of U(VI) removal from groundwater using biochar produced from hydrothermal carbonization. *Journal of Environmental Management*, 92, 2504–2512.

Lee, Y., Um, I.H. & Yoon, J. (2003) Arsenic (III) oxidation by iron (VI)(ferrate) and subsequent removal of arsenic (V) by iron (III) coagulation. *Environmental Science & Technology*, 37, 5750–5756.

Li, Y., Zhihong, S. & Yang, G. (1992) Analysis of X-ray of heart areas of 30 patients with endemic fluorosis. *Endemic Diseases Bulletin*, 7, 42.

Lin, T.F. & Wu, J.K. (2001) Adsorption of arsenite and arsenate within activated alumina grains: equilibrium and kinetics. *Water Research*, 35, 2049–2057.

Liu, Y.H., Wang, Y.Q., Zhang, Z.B., Cao, X.H., Nie, W.B., Li, Q. & Hua, R. (2013) Removal of uranium from aqueous solution by a low cost and high-efficient adsorbent. *Applied Surface Science*, 273, 68–74.

Lorenzen, L., Van Deventer, J.S.J. & Landi, W.M. (1995) Factors affecting the mechanism of the adsorption of arsenic species on activated carbon. *Minerals Engineering*, 8, 557–569.

Luss, G. (1990) Cleaning membrane systems in process industries. *Proceedings of the International Congress on Membranes and Membrane Processes, 20–24 August l990, Chicago, IL, USA*.

Ma, W., Ya, F., Wang, R. & Zhao, Y. (2008) Fluoride removal from drinking water by adsorption using bone char as a biosorbent. *International Journal of Environmental Science and Technology*, 9, 59–69.

McNeill, L.S. & Edwards, M.A. (1995) Soluble arsenic removal at water treatment plants. *Journal American Water Works Association*, 87 (4), 105–113.

Meng, X., Bang, S. & Korfiatis, G. (2000) Effects of silicate, sulfate, and carbonate on arsenic removal by ferric chloride. *Water Research*, 34 (4), 1255–1261.

Mjengera, H. & Mkongo, G. (2003) Appropriate deflouridation technology for use in flourotic areas in Tanzania. *Physics and Chemistry of the Earth*, 28, 1097–1104.

Mohan, D. & Pittman Jr., C.U. (2007) Arsenic removal from water/wastewater using adsorbents – a critical review. *Journal of Hazardous Materials*, 142, 1–53.

Mondal, P., Bhowmick, S., Chatterjee, D., Figoli, A. & Bruggen, B. (2013) Remediation of inorganic arsenic in groundwater for safe water supply: a critical assessment of technological solutions. *Chemosphere*, 92 (2), 157–170.

Montaña, M., Camacho, A., Serrano, I., Devesa, R., Matia, L. & Vallés, I. (2013) Removal of radionuclides in drinking water by membrane treatment using ultrafiltration, reverse osmosis and electrodialysis reversal. *Journal of Environmental Radioactivity*, 125, 86–92.

Nordstrom, D.K. (2002) Worldwide occurrences of arsenic in groundwater. *Science*, 296, 2143–2145.

Pallier, V., Feuillade-Cathalifaud, G., Serpaud, B. & Bollinger, J.C. (2010) Effect of organic matter on arsenic removal during coagulation/flocculation treatment. *Journal of Colloid and Interface Science*, 342, 26–32.

Parab, H., Joshi, S., Shenoy, N., Verma, R., Lali, A. & Sudersanan, M. (2005) Uranium removal from aqueous solution by coir pith: equilibrium and kinetic studies. *Bioresource Technology*, 96, 1241–1248.

Pendergast, M.T.M. & Hoek, E.M.V. (2011) A review of water treatment membrane nanotechnology. *Energy & Environmental Science*, 4, 1946–1971.

Petrusevski, B., Boere, J., Shahidullah, S.M., Sharma, S.K. & Schippers, J.C. (2002) Adsorbent-based point-of-use system for arsenic removal in rural areas. *Journal of Water Supply: Research and Technology – AQUA*, 51, 135–144.

Rahman, M.A. & Hasegawa, H. (2011) Aquatic arsenic: phytoremediation using floating macrophytes. *Chemosphere*, 83, 633–646.

Ravenscroft, A. (2009) Social software, Web 2.0 and learning: status and implications of an evolving paradigm. *Journal of Computer Assisted Learning*, 25, 1–5.

Reddy, K.J., McDonald, K.J. & King, H. (2013) A novel arsenic removal process for water using cupric oxide nanoparticles. *Journal of Colloid and Interface Science*, 397, 96–102.

Shen, F., Chen, X., Gao, P. & Chen, G. (2003) Electrochemical removal of fluoride ions from industrial wastewater. *Chemical Engineering Science*, 58, 987–993.

Shen, Y.S. (1973) Study of arsenic removal from drinking water. *Journal American Water Works Association*, 8, 543–548.

Smedley, P.L. & Kinniburgh, D.G. (2002) A review of the source, behavior and distribution of arsenic in natural waters. *Applied Geochemistry*, 17 (5), 517–568.

Sorg, T.J. & Logsdon, G.S. (1978) Treatment technology to meet the interim primary drinking water regulations for inorganics. Part 2. *Journal American Water Works Association*, 7, 379–392.

Sperlich, A., Werner, A., Genz, A., Amy, G., Worch, E. & Jekel, M. (2005) Breakthrough behavior of granular ferric hydroxide (GFH) fixed-bed adsorption filters: modeling and experimental approaches. *Water Research*, 39, 1190–1198.

Sujana, M.G. & Anand, S. (2011) Fluoride removal studies from contaminated ground water by using bauxite. *Desalination*, 267, 222–227.

Tchomgui-Kamga, E., Alonzo, V., Nanseu-Njiki, C.P., Audebrand, N., Ngameni, E. & Darchen, A. (2010) Preparation and characterization of charcoals that contain dispersed aluminum oxide as adsorbents for removal of fluoride from drinking water. *Carbon*, 48, 333–343.

Tran-Ha, M.H. & Wiley, D.E. (1998) The relationship between membrane cleaning efficiency and water quality. *Journal of Membrane Science*, 145, 99–110.

USEPA (2000) Arsenic treatment technology evaluation handbook for small systems. H.A.E.C. Division, Washington, DC.

USEPA (2003) Arsenic treatment technology evaluation handbook for small systems. H.A.E.C. Division, Washington, DC.

Vatanpour, V., Madaeni, S.S., Rajabi, L., Zinadini, S. & Derakhshan, A.A. (2012) Boehmite nanoparticles as a new nanofiller for preparation of antifouling mixed matrix membranes. *Journal of Membrane Science*, 401–402, 132–143.

Velizarov, S., Crespo, G.J. & Reis, A.M. (2004) Removal of inorganic anions from drinking water supplies by membrane bio/processes. *Reviews in Environmental Science and Biotechnology*, 3, 361–380.

Villalobos-Rodriguez, R., Montero-Cabrera, M.E., Esparza-Ponce, H.E., Herrera-Peraza, E.F. & Ballinas-Casarrubias, M.L. (2012) Uranium removal from water using cellulose triacetate membranes added with activated carbon. *Applied Radiation and Isotopes*, 70, 872–881.

Vitela-Rodriguez, A.V. & Rangel-Mendez, J.R. (2013) Arsenic removal by modified activated carbons with iron hydro (oxide) nanoparticles. *Journal of Environmental Management*, 114, 225–231.

Wang, Y. & Reardon, E.J. (2001) Activation and regeneration of a soil sorbent for defluoridation of drinking water. *Applied Geochemistry*, 16, 531–539.

Westerhoff, P., Highfield, D., Badruzzaman, M. & Yoon, Y. (2005) Rapid small-scale column tests for arsenate removal in iron oxide packed bed columns. *Journal of Environmental Engineering*, 131, 262–271.

WHO (2011) *Guidelines for drinking-water quality*. 4th edition. WHO/HSE/WSH/11.03. World Health Organization, Geneva, Switzerland. Available from: http://www.who.int/en/ [accessed April 2015].

Wu, X., Zhang, Y., Dou, X. & Yang, M. (2007) Fluoride removal performance of a novel Fe-Al-Ce trimetal oxide adsorbent. *Chemosphere*, 69 (11), 1758–1764.

Zhang, G., Gao, Y., Zhang, Y. & Gu, P. (2005) Removal of fluoride from drinking water by a membrane coagulation reactor (MCR). *Desalination*, 177, 143–155.

Zhou, J., Zhao, L., Huang, Q., Zhou, R. & Li, X. (2009) Catalytic activity of Y zeolite supported CeO_2 catalysts for deep oxidation of 1, 2-dichloroethane (DCE). *Catalysis Letters*, 127, 277–284.

Zomoza, B., Tellez, C., Coronas, J., Gascon, J. & Kapteijn, F. (2013) Metal organic framework based mixed matrix membranes: an increasingly important field of research with a large application potential. *Microporous and Mesoporous Materials*, 166, 67–78.

Zou, W., Han, R., Chen, Z., Jinghua, Z., Shi, J. (2006) Kinetic study of adsorption of Cu(II) and Pb(II) from aqueous solutions using manganese oxide coated zeolite in batch mode. *Colloids and Surfaces* A: *Physicochemical and Engineering Aspects*, 279, 238–246.

Zou, W., Zhao, L. & Han, R. (2009) Removal of uranium (VI) by fixed bed ion-exchange column using natural zeolite coated with manganese oxide. *Chinese Journal of Chemical Engineering*, 17 (4), 585–593.

Part III
New trends in materials and process development